# LIQUEFIED NATURAL GAS

# Liquefied Natural Gas

## W. L. LOM

*Esso Research Centre, Esso Petroleum Co. Ltd.,*
*Abingdon, Berkshire, England*

## APPLIED SCIENCE PUBLISHERS LTD
### LONDON

CHEMISTRY

APPLIED SCIENCE PUBLISHERS LTD
RIPPLE ROAD, BARKING, ESSEX, ENGLAND

ISBN: 0 85334 583 X

WITH 21 ILLUSTRATIONS AND 36 TABLES

Printed in Great Britain by Bell and Bain Limited, Glasgow

# Preface

Until quite recently gases produced in remote regions of the world were either considered valueless, and the wells from which they had been produced were plugged and abandoned, or if natural gas production was associated with that of oil, the 'associated' gas was either re-injected into the oil-producing formation or it was flared. Clearly the burning off of large volumes of gas, however unavoidable, was frequently considered unacceptable by local administrations and alternative uses had to be found for the associated gas. Pressure was exerted on the producing companies to convert the gas into petrochemicals and later, when this became technically feasible, to initiate gas liquefaction and LNG shipment projects. In other words, initially such projects were adopted largely to appease local objections to the flaring of large volumes of surplus gas.

More recently the attitude of oil and gas producers underwent a distinct change. Concern about air pollution in industrial countries due to the emission of sulphur oxides, carbon monoxide, other partial combustion products and particulates resulted in a clear-cut price premium for clean gaseous fuels over liquid fuels which contained some sulphur and did not burn with the same degree of cleanliness. The shipment of liquefied methane from remote gas fields to industrialised countries thus became a commercial proposition in its own right, and a number of such schemes were consequently initiated by oil and gas companies themselves without any pressure by host governments.

The most recent development involved, if not a complete change in attitude, at least one of emphasis. While the first LNG projects were initiated as much to provide clean fuels to cities polluted by smoke and other forms of emission as to mitigate shortages of locally produced gas, we note now that LNG is being imported and will be even more so in the near future mainly to supplement local gas resources which are no longer adequate. Not only is gas demand increasing very rapidly in all the industrialised countries but, as a result of this large demand, local gas sources are being exhausted at an unexpectedly rapid rate; LNG is therefore being called upon to meet energy shortages in general and local gas shortages in particular.

v

It follows that it would be difficult if not impossible to visualise circumstances which could reverse the present trend of rapidly spreading, almost proliferating, LNG projects all over the world, and of a growing use of this new fuel. It would therefore seem that this was the moment to review the status of LNG development and while doing so also summarise the state of the art in regard to gas treatment, liquefaction, transport, storage, regasification, distribution and utilisation of LNG. Such a technological review seemed the more appropriate because, at least in my opinion, the technology of liquefaction and of the handling of the resultant liquid has reached a stage where initial difficulties have been overcome and a reasonably clear-cut technical solution—or maybe several such solutions—can be envisaged for most of the problems which have been encountered in the recent past.

The present book is an attempt to deal collectively with the previously mentioned numerous aspects of LNG technology. The number of publications, papers, reports, presentations, conferences and trade literature dealing exclusively with LNG or jointly with LNG and some other topics is already very large and the material which should be reviewed is growing all the time. This presents a review author not only with a problem of selecting suitable, and rejecting unwanted, material, but also with the difficult task of selecting a suitable cut-off point as far as consideration of the technical literature is concerned.

The material reported in the book is indicative of my personal views. Although associated for the past 23 years with the Esso or Exxon group of companies, and therefore no doubt much influenced in my thinking by that of my colleagues in Esso Petroleum Company and its parent, Exxon Corporation, the views and statements expressed here are my own and not necessarily identical with the ideas and attitudes of my employers.

Nevertheless, I should like to take this opportunity to thank my colleagues at the Abingdon Research Centre and elsewhere for the opportunity to discuss my material with them and to thank the management for encouragement and support. Finally, I should like to add a word of thanks and apology to my long-suffering wife and family without whose forbearance this book would not have been written.

W. L. LOM

# Contents

## Chapter 1

# Introduction

Over the last ten years the use of a new fuel, or perhaps more accurately, of an old fuel in a new form, has progressed at an extraordinarily rapid rate, and has made its first impact not in just one but in many regions of the world. The fuel in question is natural gas, a hydrocarbon or fossil fuel consisting mainly of methane, and its relatively new form is liquefied natural gas, usually referred to as LNG.

The reasons for the sudden spread of LNG production schemes, both projected and already in existence, are manifold and are closely linked to the economics of energy generation, the growing concern about air pollution and conservation of the environment, the necessity of a fuller utilisation of natural resources, the objections raised by the governments of oil-producing countries against the flaring of by-product gas from crude oil operations, and last but not least, the need to supply the ever-growing demand for energy by both developed and developing regions around the globe.

The fact that natural gas is not as a rule produced in close proximity to regions of high gas demand, and that in those instances where gas fields occur in industrialised areas they tend to be exhausted fairly rapidly, has resulted in many parts of the world in major movements of gas by pipeline from gas fields to the centres of consumption. The gas, after purification and conditioning, is compressed and transported over large distances. Pressures of 100 bar and pipelines of anything up to several thousand kilometres long are in use.

However, pipelines and piped gas transport have a number of drawbacks. Amongst others, pipelines cannot at present be laid at depths more than about 100 m below sea level, and in any case underwater pipelines are expensive and sometimes at risk. Political conditions in the areas between producer country and the centres of gas consumption may not be conducive to investment, and gas supplier and client may not wish to expose a relatively vulnerable link to the vagaries of unstable political systems. Furthermore, pipelines are permanent fixtures leading from a given gas field to a particular

1

consuming area, whereas both sources of gas and concentrations of gas consumption are subject to change. Finally, gas stored in a pipeline or under pressure can only be used to a limited extent to follow a fluctuating gas demand by allowing pressure changes in the line, a procedure known as line-pack.

Liquefied natural gas, on the other hand, is ideally transported across the sea, although road, rail and particularly barge movements are operational in various parts of the world. Transport from field to consumer is therefore largely across the open sea and unhampered by political instability en route. And while liquefaction plants and loading facilities are permanently tied to a particular gas field, the LNG ships, which represent the largest investment in an LNG scheme, can be moved from one supply system to another, and as the number of gas liquefaction plants around the world increases this extra flexibility is becoming of real value. Similarly, destination of the LNG, once it is loaded, can be changed in accordance with demand; a particular instance which springs to mind is seasonal gas demand in the Northern and Southern hemispheres. Finally, LNG, once liquefied and refrigerated, can be stored at atmospheric pressure and therefore constitutes a useful gas reserve to be drawn upon when demand temporarily rises above average.

While these advantages of an LNG supply over a piped natural gas system have been well known for many years and the basic technology of liquefaction has also been established for some time now, the introduction of LNG as a bulk energy source is relatively recent and is due, in addition to the economic and environmental factors just mentioned, to a number of technological developments. These can be subdivided into: first, technical progress associated with the liquefaction process, including gas purification before lique-faction and gas separation: secondly, the technology of constructing LNG ships, particularly large tankers for long distance transport, and including such aspects as loading and unloading from single-point mooring buoys: and thirdly, the design and construction of cryogenic storage tanks using new materials, new methods and improved insulation.

Finally, there is little doubt that recent studies of hazards associated with LNG production, transportation and storage have resulted in improved safety features of LNG schemes all the way from the liquefaction plant to the ultimate consumer.

Apart from technological changes in LNG production, shipment, storage and safety, the availability of LNG has itself generated new technology in the shape of novel methods of LNG transport from storage to consumer and direct use as fuel or cryogenic fluid by consumers, occasionally without prior regasification. Its availability has also led to the extension of established LNG uses, e.g. for peak shaving in gas distribution systems, and to fresh thinking about the supply of clean energy to domestic and commercial consumers.

Any survey of the LNG field therefore, in addition to discussing the history of gas liquefaction processes and projects and reviewing briefly the technology and economics of natural gas production, must focus attention on the topics of natural gas liquefaction technology, the loading and shipping of LNG, its storage and regasification, the hazards associated with its handling and the safety precautions required, and finally its distribution and utilisation.

Although most technical papers dealing with LNG are relatively recent, their number already goes into four figures. Specialisation in particular aspects of LNG technology is becoming well established; the present study is an attempt to pull together information bearing on LNG without any hope of satisfying the specialist in a particular aspect of its technology. Coverage of topics, size of the work and the time available for writing have forced us to concentrate on essentials; for detailed information the reader is referred to the numerous technical papers and articles quoted in the literature reviews at the end of each chapter.

*Chapter 2*

# Historical Review

## 2.1 ORIGINS OF CRYOGENICS

The science of cryogenics or generation of low temperatures is the outcome of a better understanding of phase relationships and of the energy and heat transfers involved in phase changes, which resulted from the studies of Gibbs, Carnot and others in the late eighteenth and early nineteenth centuries (references to the various 'classical' publications mentioned are listed in Ref. 14). It may be of some significance that the actual achievement of the low temperatures required for the liquefaction of the so-called permanent gases did not take place until about fifty years after the establishment of the thermodynamic principles involved. The first successful liquefaction of permanent gases on a laboratory scale was carried out by Michael Faraday in 1845. Subsequent pioneering work on gas liquefaction techniques by Louis Cailletet, Raoul Pictet, Olszewski, Dewar and Kamerlingh Onnes in France, Switzerland, Poland, England and Holland respectively, enabled Carl von Linde in Germany to set up in 1895 the first continuously operating air separation plant based on the liquefaction of air. All modern industrial gas liquefaction processes can thus be said to be derived from Linde's work on the air liquefaction cycle.

It is proposed to discuss thermodynamic cycles used for the refrigeration of natural gas in greater detail in Chapter 3 under the heading of Gas Liquefaction Plants and in Appendix A. Suffice it to say at this stage, that the original Linde air liquefier operated on the basis of isenthalpic or adiabatic expansion of compressed air through a reducing valve. The resultant cooling of the gas is known as the Joule–Thompson effect. The Linde cycle thus differs from the other important refrigeration cycle, the Claude liquefaction system, which was developed a few years later (1902) and uses mainly the isentropic expansion of a gas in a gas engine to extract work from the system during expansion. More recently (1937) the original Claude reciprocating expansion engine has been replaced by an expansion turbine.

In neither case, however, is isentropic expansion the exclusive cooling mechanism; since piston and bearing lubrication present problems at low temperatures, final cooling and liquefaction are usually achieved by isenthalpic expansion. In the Heylandt liquefier, which was developed in the 1920s, about half the refrigeration is due to isentropic expansion and the remainder is achieved by release of the gas through a throttle valve. In the earlier Claude process about 20% of the cooling effect is due to isenthalpic expansion and 80% to expansion in an engine.

In all these instances it is impossible to reach gas liquefaction temperatures in a single expansion stage and it is essential to recompress the gas, cool and expand it again a number of times. It follows that the cold gas after expansion has to be heat-exchanged with hot compressed gas to precool the latter before expanding it. Efficient heat exchange and the design of suitable gas-to-gas exchangers are therefore part and parcel of a successful gas liquefaction system. It is another of Linde's firsts that his continuous air liquefaction plant incorporated an efficient heat exchanger.

Cryogenic heat exchangers differ from plant used at room temperatures and above in several important respects. The rate of heat transfer at the lower temperatures used is slow; nevertheless, total heat energy transferred in the course of liquefying a gas is substantial. As a result the area and the total length of tubular exchangers used in cryogenic processes are invariably very large; particularly tube-in-tube exchangers, with a number of thin tubes encased in a helical coil of larger tubing, tend to become unmanageable. Cross-counter-current exchangers with a number of small tubes in parallel and exposed to a cross current of the heat exchange gas are shorter, but still require a helically wound arrangement. Both types present problems when the gas contains impurities that form solid deposits on freezing, e.g. water, carbon dioxide or higher hydrocarbons.

Regenerators, first introduced by Fränkl in 1928, can overcome this problem. Since they operate in tandem, one vessel packed with high heat capacity material is cooled by a gas stream while the other is reducing the temperature of the incoming gas. By arranging flow of the two gases in opposite directions, deposition of some impurities becomes permissible since they are removed to some extent by sublimation in the next phase of the cycle.

While the production of very pure liquefied gases invariably involves the use of heat exchangers, regenerators lend themselves to the bulk cooling of impure gases and to heat exchange between gas streams of similar volume and pressure differing only in temperature.

A further development in gas liquefaction which also dates back to the 1920s is the observation that the isenthalpic expansion of a gas becomes more efficient, i.e. more nearly reversible, the lower the temperature. Generally, close to the liquefaction point Joule–Thompson expansion is almost reversible and the entropy gain, even for a large pressure drop, is small. It follows that it will be more efficient to liquefy a higher boiling gas and use it to refrigerate a lower boiling gas before expanding the latter, rather than to cool the lower boiling gas by multi-step expansion with intermediate cooling.

The resultant 'Cascade' process in which, for example, ammonia is compressed and liquefied and the latent heat of vaporisation of ammonia vapour is used to precool compressed ethylene before expansion, and finally ethylene, cooled and liquefied by expansion, precools methane, is consequently more efficient than the direct expansion of methane, even if performed in several stages with heat exchange and intercooling. The concept of cascade refrigeration and liquefaction goes back to Pictet, but the first working model was built by Kamerlingh Onnes.

While many modern gas liquefaction plants still use the classical cascade process—several very large cascade plants have been built in Alaska and Algeria for instance—the use of pure refrigerants used in separate refrigeration cycles has now been largely superseded by different versions of the modified or autorefrigerated cascade process in which a mixed refrigerant stream undergoes stepwise condensation and gradual revaporisation.

First proposed by Haselden[10] for use in LNG plants, the concept was taken up by Kleemenko,[11] who built the first laboratory demonstration plant, and was incorporated in a full scale natural gas liquefaction plant by the French Technip company, whose scientific director Perrel described four possible variants[13] using two pressures. The first single-pressure suction system for natural gas liquefaction using a mixed refrigerant is due to Air Products and was not published until 1971.[6]

## 2.2 GAS SEPARATION PROCESSES

The original purpose of practically all attempts at gas liquefaction was the subsequent separation of the individual components of the liquefied gas and the preparation of pure gases.

Linde's work in Germany was not only applied to air and the preparation of pure oxygen, which was required for chemical and metallurgical uses, cutting, welding, the gasification of coal, etc. It was also used to separate hydrogen from coal gas in order to produce ammonia during the first world war and was developed to assist in the manufacture of synthetic liquid fuels between the wars. Subsequently ethylene was separated from the residual gas fraction and used for chemical synthesis.

Refinery gases and natural gas condensate were split into their components using cryogenic separation processes, and the development of the US LPG industry is due to improvements in cryogenic techniques in the 1930s. The application of these techniques to cracked gases subsequently formed the basis of the world-wide growth of the petrochemicals industry, i.e. the utilisation after their separation from other cracker gases and from each other, of ethylene, propylene and other olefins for further processing and conversion into chemicals.

Apart from producing oxygen, the numerous air separation plants which were built from 1920 onwards could also be used to recover the minor components of atmospheric air, mainly the inert gases argon and krypton, both employed in the manufacture of light bulbs.

The interest in lighter-than-air air transport, originally based on hydrogen, which was soon found to be dangerously inflammable, focussed attention on helium, a minor component of certain natural gases. In order to separate helium from the remaining natural gas components it was necessary to refrigerate and condense the bulk of the gas, and the first liquid methane was obtained as a by-product of helium separation in 1924 by the US Bureau of Mines. Interest in helium for airships has since abated, but the gas is used as a heat transfer agent in heavy water and high temperature gas-cooled nuclear reactors, and finds certain other applications in the generation of atomic energy. The liquefaction of natural gas purely for helium recovery is now confined to certain helium-rich gas fields in the United States and in Russia. The techniques developed for helium recovery, however, have since been refined and applied to the liquefaction of natural gases for subsequent movement by ship, road and rail.[1]

## 2.3 THE UTILISATION OF COLD

Apart from gas separation, liquefied gases also find an important outlet in all forms of refrigeration. While complete refrigeration cycles are based on the compression and re-evaporation of certain gases, particularly ammonia but also propane and ethylene, other liquefied gases can be made to absorb heat before being utilised in the vapour or gaseous form. Liquid nitrogen, for example, is frequently used as a source of extreme cold. Solid carbon dioxide, the liquid form being unstable at atmospheric pressure, is the commonest source of milder refrigeration. The achievement of extremely low temperatures, close to absolute zero, e.g. to achieve superconductivity in metals, requires liquid or boiling helium.

The combination of gas separation processes, mainly the production of oxygen, with refrigeration by means of cold or liquefied gases is not uncommon. Food, and particularly meat, storage in the Chicago area uses cold nitrogen from oxygen plants which supply nearby steel mills; the first experiments with liquid methane transport were carried out by the Chicago Union Stockyards, who intended to use the cold produced by re-evaporating the liquid.

The 'cold' generated during LNG revaporisation can also be returned to the natural gas liquefaction plant. Air or nitrogen is liquefied at the point of natural gas use and can be shipped back to the source of the LNG, where it is vaporised to precool the incoming natural gas before liquefaction.

Other recent suggestions to use liquefied gases for refrigeration include such concepts as the cooling of supersonic aircraft by LNG, which is subsequently used as fuel, the refrigeration of trucks propelled by LPG or LNG by means of boiling fuel, and the combined shipment of perishable goods and liquefied gases.

## 2.4 CRYOGENICS IN THE GAS SUPPLY INDUSTRY

The supply of piped gas to industry, households, commerce etc., is one of the features of civilised societies. Both natural gas and gaseous fuels produced from solid or liquid hydrocarbons have been distributed by pipeline for more than a century in many parts of the world.

However, one of the basic problems of a piped fuel supply is the unsatisfactory utilisation of expensive supply facilities due to daily, weekly or seasonal variations in demand. Meeting sudden peak requirements, without excessive investment in equipment which would be used only on rare occasions, is essential to the industry. And since it is extremely expensive to store gas as such in large quantities at moderate pressure many ingenious attempts have been made in the past to develop peak gas generating facilities that can be based either on gas manufacture or on gas storage.

The latter, to become economical, had to be at elevated pressure, and both pressurised underground reservoirs and above ground or in-ground pressure vessels are in use. However, even compression to 1000 psig (70 atm) reduces the volume of a gas only by a factor of 70, whereas liquefaction of methane results in a volume contraction of 580 to 1, a ratio which allows one to store sufficient gas to meet peaks of demand by means of relatively small storage vessels.

The use of liquefied natural gas for distribution by the gas industry was first proposed in the United States in 1930; however, the first liquefaction plant for peak shaving was not built until 1940. Further construction of peak shaving facilities followed until in 1944 a major accident and disastrous fire at the Cleveland LNG storage plant set back LNG peak shaving developments by a number of years. No further commercial plants were, in fact, built in the US until the early sixties, but a large liquefaction plant was supplied by the United States to Russia under Lease-and-Lend and was installed in Moscow in 1947.

A combined exercise by Union Stock Yards and Continental Oil (Constock) in 1952 envisaged barge shipment of liquefied gas from a barge-mounted liquefaction plant on Lake Charles, Louisiana, up the Mississippi to Chicago; however, the project was abandoned on safety grounds and the first major transport of LNG had to wait until 1959 when a modified crude oil tanker, the *Methane Pioneer*, carried several cargoes of LNG from Lake Charles, Louisiana, to Canvey Island in the Thames estuary.

TABLE 2.1

Natural Gas Liquefaction Plants for Peak Load Shaving

(Totals only, detailed list in Appendix F)

| Country | Number of schemes | | Storage capacity ($10^6$ scfd) | |
|---|---|---|---|---|
| | Existing | Proposed | Present | Future |
| United States | 45[a] | several | 45 700[a] | ? |
| United Kingdom | 3 | 4 | 3 600[b] | 9 500[b] |
| Canada | 3 | 2 | 2 800 | ? |
| Germany | 1 | — | 770 | 770 |
| Netherlands | — | 1 | — | 2 000 |

[a] Satellite plants without liquefaction facilities or base load plants not included.
[b] Does not include storage at Canvey for imported (base-load) LNG.

Simultaneously, new methods of LNG tank construction were developed: these included so-called frozen holes, where, as the name implies, ice formed from ground moisture contains the liquid. An insulated cover completes the vessel. Alternatively, prestressed concrete tanks, sunk into the ground, were found acceptable and improved forms of above-ground tankage, both concrete and with double metal walls, were constructed.

As a result the storage of LNG for peak load shaving has by now become an established means of evening out load fluctuations, and a large number of liquefaction plants in the United States, Europe/ UK and Russia are now in operation.[2,9]

In addition to the use of LNG for peak shaving, the transport of liquefied gas from liquefaction plants to consumers to meet both peak and base gas demand has developed extensively since the early days of the *Methane Pioneer*. Numerous LNG projects involving

collection of natural gas, transport by pipeline to a coastal liquefaction plant, liquefaction and loading facilities, ships designed
from the start to carry LNG (rather than converted tankers),
unloading and liquid storage plant at the consuming end, and, finally,
regasification facilities, have been built over the last few years
to serve the gas industries of many countries.

The gas distribution industry uses LNG in two somewhat different
ways. It liquefies surplus gas delivered during the summer season
and stores it for re-injection into the system during periods of high
gas demand. Table 2.1 lists the number of schemes of this type in

TABLE 2.2

**Gas Liquefaction Plants for LNG Export**

(More detail in Appendix F)

| Country | Number of facilities[a] | | Liquefaction capacity, ($10^6$ scfd) | |
|---|---|---|---|---|
| | Existing | Proposed | Existing | Proposed |
| Algeria | 2 | 3 or 4 | 600 | 4 000 |
| Alaska | 1 | 1 | 90 | 250–1 000 |
| Libya | 1 | — | 345 | None |
| Brunei/Sarawak | 1 | 1 | 500 | 1 300 |
| Abu Dhabi/Qatar | — | 1[b] | — | 600 |
| Iran | — | 3[b] | — | 2 000–2 800 |
| Venezuela/Trinidad | — | 2 | — | 800–1 500 |
| W. Australia/Cent. Australia | — | 2 | — | 2 300–3 000 |
| USSR Baltic/E. Siberia | — | 2[b] | — | 3 000 |
| Nigeria | — | 3[b] | — | 1 500 |
| Indonesia | — | 2[b] | — | 2 000 |
| Other | — | 3[b] | — | 1 000+ |
| Total | 5 | 24 | 1 535 | 18 750–21 700 |

[a] Liquefaction plant locations: there are invariably several liquefaction trains and
in some instances more than one liquefaction plant at any one export terminal.
[b] Some doubt whether all the projects listed will come to fruition.

different countries which have been reported over the past few years
and their capacity; a detailed list will be found in Appendix F.
Alternatively it specifically produces and collects gas for liquefaction
and transports it in liquefied form to the consuming area. The latter
schemes are, of course, much more involved, considerably more
expensive and therefore at present fewer in number. However, as will
be gathered from Table 2.2, by now five liquefaction schemes of this
type are operational and a large number of similar projects are being
built, planned, or considered at the time of writing.[3,4,7]

The liquefied gas will be delivered to terminals in areas of excess
demand, mainly in the US, Japan and Europe, and Table 2.3 lists

import schemes by country and in terms of total capacity in the years 1975, 1980 and 1985.[8,12,15] While volumes of gas liquefied, on the one hand, and imported into LNG terminals, on the other, are the same, the number of projects in hand for terminals and liquefaction plants may be different, since more than one LNG plant may supply any one terminal, and gas from one liquefaction plant will often be shipped to several destinations. Details of individual LNG schemes are listed in Appendix F.

TABLE 2.3

Major LNG Import Schemes

(Details in Appendix F)

| Country | Number of schemes[a] | | Expected LNG volumes ($10^6$ scfd) | | |
|---|---|---|---|---|---|
| | Existing | Proposed | 1975 | 1980 | 1985 |
| United States | 1 | 15 | 120 | 4 500 | 11 000 |
| Japan | 2 | 12 | 935 | 4 600 | 7 700 |
| Europe: UK | 1 | — | 100 | 100 | 100 |
| France[b] | 2 | 1 | 400 | 650 | 750 |
| Italy[b] | 1 | 1 | 235 | 235 | 235 |
| Spain | 1 | 1 | 110 | 260 | 260 |
| Belgium | — | 1[c] | — | 160 | 250 |
| Germany | — | 1[b] | — | 590 | 700 |
| Other | — | 2 | — | 205 | 205 |
| Total Europe | 5 | 7 | 845 | 2 200 | 2 500 |
| World total | 8 | 32 | 1 900 | 11 300 | 21 200 |

[a] Total number of import schemes is not the same as that of LNG terminals, since LNG may be unloaded in more than one port, or a terminal may receive imports from several sources.

[b] Algerian LNG for southern Germany will be imported via both Fos-sur-Mer and Trieste.

[c] LNG may not reach Belgium and Germany via terminals in Southern Europe.

The large quantities of natural gas that are to be liquefied in areas of surplus availability and will be consumed by industrial societies in high population density areas elsewhere require not only liquefaction and revaporisation facilities at each end of the voyage but also a large number of ships to carry the liquefied gas cargo. Apart from the technological developments over the last few years, which have led to safer and more economic LNG carrier construction, the sheer number of ships completed, building and projected must be considered a remarkable technological achievement in its own right. The list of ships in Appendix G is an indication of progress in this area over recent years.[5]

12                    LIQUEFIED NATURAL GAS

# REFERENCES

1. American Gas Association (1968). *LNG Information Book*, New York.
2. Anon (1972). LNG scoreboard, *Pipeline Gas J.*, **199**(7), 44–46.
3. Anon (1972). LNG 1971: profile of progress, *Cryog. Ind. Gases*, **7**(1), 25–27.
4. Cranfield, J. (1971). Vast Brunei LNG scheme is right on schedule, *Oil Gas Int.*, **11**(7), 24–26.
5. Faridany, E. (1972). *Marine Operations and Market Prospects for Liquefied Natural Gas* 1972/1990 QER Special No. 12, The Economist Intelligence Unit Ltd., London.
6. Gammer, L. S. and Newton, C. J. (1971). US Patent 3,593,535.
7. Goodwin, A. M. (1971). Fos LNG terminal will be operational early in 1972, *World Pet.*, **42**, 34–36, Aug.
8. Hale, D. (1972). LNG continues spectacular growth, *Pipeline Gas J.*, **199**(7), 41–43.
9. Hanke, C. C. and Lizinger, L. F. (1971). Baltimore's new plant now in operation, *Pipeline Gas J.*, **198**(12), 64–70.
10. Haselden, G. C. and Barber, N. R. (1957). *Trans. Inst. Chem. Engs.*, **35**, 2, 77.
11. Kleemenko, A. P. (1960). One flow cascade cycle, in *Progress in Refrigeration Science and Technology* (eds. M. Jul and A. M. S. Jul), Vol. 1, p. 34, Pergamon Press, Oxford.
12. Kroeger, C. V. (1972). LNG on the global scene, *Gas*, **48**(7), 38–41.
13. Perret, J. (1966). Techniques in the liquefaction of natural gas, *Petrol. Times*, Nov.
14. Ruhemann, M. (1949). *The Separation of Gases*, 2nd ed., Clarendon Press, Oxford.
15. Smith, W. H. and Anderson, P. J. (1972). LNG imports to the US—Here's where they stand today, *Pipeline Gas J.*, **199**(7), 91–95.

# Chapter 3

# Natural Gas Supply and Demand

## 3.1 INTRODUCTION

In order to appreciate the significance of LNG to the gas distribution industry it is first necessary to review the sources of supply and the pattern of sales of a typical gas distribution company. From the resultant supply/demand balance there should develop an indication where LNG could be used to some advantage. It has, however, been mentioned that the LNG route is basically much more complex and also more expensive than the direct linking of gas source and consumer by a system of pipelines; the aim of this discussion is to establish conditions under which LNG is attractive in spite of these handicaps.[4]

Although only a few years ago any analysis of gas distribution would have had to differentiate clearly between the situation in the United States and that in Europe, most of the differences which have been evident for many years are in the process of disappearing. In particular the widespread distribution of manufactured gas in Europe—originally made from coal, more recently from oil—has given way to natural gas in the major European markets. In line with natural gas penetration the separate and independent gas grids of the past have ceased to exist and a gas of uniform quality is piped over considerable distances from the gas fields in North Holland, South Western France, Northern Italy and off-shore in the North Sea to the centres of population and industrial production. Piped gas imports over long distances from Russia have been initiated and the construction of pipelines from North Africa is being considered.

This situation is not too different from that in the US where domestic production in Texas, Louisiana, Oklahoma and other States with surplus production has been piped to the industrial North East, and piped supplies from Canada have been brought in to augment domestic sources. Very similar again is the gradual depletion of the older fields and a developing shortage of domestic gas which has resulted in a number of major import schemes.

Another point of similarity is the extremely seasonal character of gas demand. While not so long ago domestic heating was fairly rare in Europe, the convenience of a piped supply, the cleanliness of gas combustion and a rising living standard have resulted in a similar consumption pattern to that in North America.

Even in industrial sales of natural gas Europe seems to follow in the steps of the US. Unsatisfactory load factors, i.e. insufficient overall utilisation of pipelines and other plant due to seasonality of demand, have driven the gas distributors both in the US and in Europe into major gas sales, at relatively low prices, to industrial consumers whose demand changes little with the seasons. Thus bulk sales to power plants and to the chemical, steel, ceramic and glass industries have helped to operate the new gas wells at a constant rate and to fill the new pipelines at an early date.

Not altogether unexpectedly, on both sides of the Atlantic, sooner or later a shortage of gas for essential uses and an inability to supply the extremes of demand on winter peak days from available gas sources have subsequently developed. But while the shortage in the US is already calling for drastic remedies it seems that European supplies, particularly if North Sea gas is fully exploited, will last a little longer.

In either situation one can solve some of the problems of excessive gas demand by judicious use of gas liquefaction. The role of LNG in the gas industries of both the US and Europe, and also Japan, is twofold; on the one hand LNG can help to even out the fluctuations of gas demand due to changes in temperature and industrial activity; on the other LNG schemes can be used to meet developing shortages of locally produced or imported gas.

## 3.2 THE CONCEPT OF PEAK SHAVING

The demand pattern of a gas distribution system serving a domestic market in medium and high geographic latitudes is shown diagrammatically in Fig. 3.1. This plots the gas demand of the complete system against the number of days in the year in which it is equal or higher than a particular figure. It will be noted that a relatively low base requirement for cooking and water heating extends throughout the year, whereas the winter heating load, reinforced by additional hot water demand, results in a fairly steep curve with a peak value corresponding to gas consumption on the coldest day of the year.

A distributor whose market is entirely domestic will thus be supplying anything between ten and fifty times the volume of gas on the coldest winter day, compared with the minimum consumption day corresponding to the lowest point of the curve. His facilities will have to be designed to cope with this situation and will, therefore, be under-utilised during much of the year, i.e. his load factor, defined as send out volume/send out capacity, will be low. Furthermore, if he buys gas from a pipeline in accordance with his consumption pattern, the pipeline will also have to maintain sufficient spare capacity which will again be under-utilised, and his gas tariff will be correspondingly unfavourable.

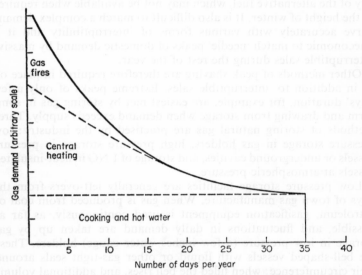

FIG. 3.1 Simplified gas demand/duration curve—domestic market.

There are ways of improving this situation; the inclusion of a commercial or industrial gas load, which does not fluctuate seasonally, will lessen the ratio of maximum to minimum day demand and will result in improved plant utilisation, i.e. will raise load factor. If overall gas demand is growing, a rapid build-up is usually easier in industry than in domestic or commercial sales and bulk gas sales are often used as a stop gap.[4] But the most useful means of eliminating or reducing seasonality is to develop outlets which can be controlled, i.e. are effectively non-seasonal.

### 3.2.1 Interruptible Gas Sales and Gas Storage
The most effective short-term means of matching capacity with demand are so-called interruptible sales. Interruptible gas is supplied to industrial customers, e.g. to power stations, during the summer

months when domestic gas demand is low. In the winter the gas is sold to households and an alternative fuel is used for steam raising or power generation.

While interruptible gas sales are very effective in improving system load factors, the prices obtainable from industry are invariably low, since interruptible gas customers not only have to provide facilities for dual firing, but frequently also have to store a substantial quantity of the alternative fuel, which may not be available when required at the height of winter. It is also difficult to match a complex demand curve accurately with various forms of interruptibility and it is uneconomic to match 'needle' peaks of domestic demand by massive interruptible sales during the rest of the year.

Other methods of peak shaving are therefore required in place of, or in addition to, interruptible sales. Extreme peaks of only a few days' duration, for example, are easiest met by storing gas in some form and drawing from storage when demand exceeds supply. Three methods of storing natural gas are practised by the industry: low pressure storage in gas holders, high pressure storage in pressure vessels or underground cavities, and storage of LNG in heat-insulated vessels at atmospheric pressure.

Low pressure storage facilities are generally left-overs from the days of town gas manufacture. When gas is produced from coal or petroleum, gasification equipment is used continuously, as far as possible, and fluctuations in daily demand are taken up by gas storage in low pressure (a few inches water gauge) holders. These are bell-shaped vessels with liquid or other gas-tight seals around their circumference; when filled the bell rises, and additional volume is created by mounting several telescopic rings inside each other. However, the cost of low pressure gas holders, in relation to their capacity, is high and their height, large site area and unattractive appearance, reinforced by the need to expand natural gas before it enters such storage and sometimes to recompress it for distribution, have almost put an end to new construction of telescopic holders.

Pressure vessels, on the other hand, are becoming more popular these days. While it is possible to reduce pressure in pipelines used for gas transport whenever gas demand rises and to restore line pressure again during subsequent periods of lower demand, the use of this so-called line-pack reduces the overall carrying capacity of a system, which is clearly highest if pressures are constantly maintained near their maximum. It is not uncommon, therefore, to install high pressure gas holders near the centres of gas consumption in the form of above-ground spheres, or bullet shaped vessels, buried large diameter pipe or bullets or, finally to inject gas into underground formations suitable for gas storage.

The latter usually require the existence of subterranean salt strata of suitable topography. Not only is salt impermeable to gas, but caverns can also be washed out fairly easily by water injection. While often quite cheap in relation to the large volumes of gas which can be stored, underground caverns take considerable time to excavate or dissolve out of the rocks and further time to fill with pressurised gas before they can be put into use.

In addition to man-made caverns gas can also be stored underground in exhausted gas fields. Depending on location of the field, permeability of the strata in question and original methods of gas production, natural gas can be injected under pressure into old gas or oil production wells and withdrawn as required.

A decision as to the best method of gas storage for a particular situation will depend on the type of peak shaving, i.e. number of days per year and rate of production required, and on the relative cost of the alternatives. Clearly overall gas cost must be minimised, and mathematical methods of evaluating total cost of distributed gas from a given set of capital and operating costs for a number of storage alternatives have been published.[4,6]

Storage of gas in the form of LNG has an important part to play in any optimisation pattern, generally as a means of meeting needle peaks of a few days duration.[8,14] A calculated example in Appendix C shows that a typical load duration curve encountered by a US public utility could best be met by injecting LPG into the grid to meet peaks of a few hours' duration; extremes of a few days were supplied by drawing on LNG storage; any seasonal variations lasting a matter of weeks required underground storage; in the case of the utility in question this meant a supply of 'seasonal gas' withdrawn by the pipeline company from the latter's storage. The remaining requirements were met by drawing on their standard supply contract.

### 3.2.2 LNG Peak Shaving Plants

The purpose of a gas liquefaction plant for peak shaving is to reduce the volume of summer surplus gas sufficiently to be able to store the gas at reasonable cost until it is required for winter use.[7] It thus differs from base load liquefaction plants in several respects:
— the gas is already purified since it is supplied by pipeline and no further conditioning is normally required;
— gas composition is uniform, i.e. mostly methane plus, sometimes, nitrogen;
— it is not essential to liquefy the entire throughput, i.e. there is no need to recycle;
— the liquefaction side of the plant is smaller than most standard base load liquefaction plants and also handles gas at a much

lower rate than does the corresponding vaporisation equipment;

— gas vaporised from storage need not be reliquefied.

Peak liquefaction plants will consequently be simpler in design and construction than base load plants; the elaborate gas purification section of the latter will be absent or can be replaced by a simpler system, often consisting of molecular sieve absorbers.

Furthermore, peak liquefaction facilities can be simplified compared with LNG base load plants with a crude natural gas feed since feed gas composition will not vary, and most types of liquefaction cycle will therefore be applicable.

Finally, it may be possible to combine gas pressure reduction with liquefaction; all the pipeline gas arrives under pressure and is expanded before local distribution but only a small proportion is liquefied and stored for winter use, i.e. there is some surplus energy available for liquefaction.

A point to bear in mind is that liquefaction—and similarly revaporisation—of a gas mixture will result in a change of composition, higher molecular weight components being preferentially condensed and also retained in liquid form on evaporation. If constant quality of peak load send-out gas is important it may be necessary to fractionate gas blends in the course of condensation, and in at least one instance this is done. In one European peak gas liquefaction plant Dutch natural gas is fractionally condensed to rid it of its nitrogen content and a pure methane fraction is stored for winter peak shaving. In less critical situations, on the other hand, slight fluctuations in composition and gas combustion properties between arriving and send-out gas may be acceptable.

Design and construction of LNG storage tanks in peak shaving and base load liquefaction plants are not too different, although the ratio of storage volume to daily liquefaction capacity is clearly very much higher for peak load facilities (between 100 and 300) than for base load plants (between 4 and 6). Furthermore, peak load plants are often built in or near densely populated areas and in consequence have to incorporate much more elaborate safety precautions.

Finally, peak load plants have provision for revaporisation of the liquefied gas. A number of possible alternatives for this stage will be discussed in Chapter 7 in greater detail. Heating of vaporisers can be by a direct firing, through either a heat exchanger, or a submerged combustion system. It can also be indirect by heating an intermediate fluid such as steam or water. In certain instances sea water is circulated through large heat exchangers. Vaporisation capacity is generally much larger than liquefaction capacity and in extreme cases the whole of the LNG in store could be revaporised in as little as four days continuous operation. However, since LNG is normally used to meet needle peak demand only one would not expect LNG peak shaving facilities to be in operation for more than a few hours at a time.[13]

A particular form of peak shaving plant is the so-called satellite LNG facility. This generally consists of a storage tank, a vaporiser and odorisation equipment. The tank is filled by road, rail or barge from a much larger peak shaving LNG plant, and clearly several satellites can be supplied from one central liquefaction plant.

Satellite LNG plants usually operate unattended, the flow of LNG to the vaporiser being regulated by the gas pressure in the distribution grid. Other advantages of satellites over central LNG facilities are the reduction in gas volume which has to be sent out from the central plant at peak gas consumption times and their rapid response to increased local gas demand i.e. minimum pressure fluctuations in the grid.

In addition to their function as peak gas producers, satellites can also be used to distribute gas in new areas which are not, or not yet, connected to the main supply system. Under these circumstances a local LNG tank is filled regularly from the central tankage throughout the year; savings result mainly from not having to build a high or medium pressure gas main from the central liquefaction plant to the satellite area.

## 3.3 LNG FOR BASE LOAD GAS SUPPLIES

While gas liquefaction to meet periodic temporary gas shortages from equally periodically occurring gas surpluses has an inherent logic, the supply of base load gas through liquefaction only becomes acceptable under certain conditions. In particular, either the supply of gas by pipeline must be insufficient to meet demand, e.g. owing to depletion of the fields, or it could be unreliable or possibly too expensive.[9] It is, however, axiomatic that the transport of gas from point A to point B over land will invariably be cheaper by pipeline than by liquefaction, followed by sea transport, and ultimate regasification. Most base load LNG liquefaction schemes will, therefore, be associated with LNG shipment by sea.

For an economically attractive LNG scheme it will furthermore be essential to have a cheap source of natural gas for which there is little or no local outlet. In order to assess the possibility of natural gas liquefaction for shipment overseas it will, therefore, be necessary to review present and future gas demands in relation to gas reserves in different parts of the globe.[12]

### 3.3.1 Areas of Gas Surplus and Deficiency

In the past natural gas has generally been found as a by-product of the exploration for oil. Not only are geological conditions for oil and gas accumulation not dissimilar but most oil fields also produce substantial quantities of so-called 'associated gas'. In consequence surplus gas is widely available where large oil discoveries have been made in the past e.g. Iran, Iraq, Kuwait, Saudi Arabia, the smaller Arabian Gulf countries, Libya, Venezuela, Nigeria, Indonesia and Alaska. The bulk of the gas co-produced with oil in these countries is stripped of condensible components and either flared or compressed and re-injected into the formation to maintain pressure and facilitate further oil production. The flaring of surplus gas is clearly objectionable on a number of grounds and alternative uses for such gas are constantly being sought.

In addition to associated gas, substantial discoveries of dry gas have been made over the years, on the one hand, in the search for oil, on the other while actually looking for gas.

Instances of the former are the discovery of gas fields in the United States and Canada, Algeria, USSR, Iran, Venezuela, Holland, Borneo and Australia. The search for natural gas as such has radiated in the past from the centres of consumption, e.g. in North America from Pennsylvania outwards towards the Southern US and Canada, in the USSR from the Ukraine towards the Caspian, the Urals and eventually Siberia. In Europe discoveries in Germany were followed by those in Holland, the southern parts of the North Sea and eventually Norwegian waters. In addition there are relatively isolated gas discoveries such as those in Western Australia and New Zealand, in and around the Adriatic Sea and in Pakistan and elsewhere.

Table 3.1 lists reserves of the gas fields mentioned above in terms of regions of the world. It will be gathered that the largest untapped reserves are in Central Asia and the Middle East with other substantial volumes available from North Africa, South America, Indonesia and Australia.[11,12]

Natural gas exporting countries with unusually large surpluses are:[11] Algeria ($3.7 \times 10^{12}$ m$^3$), Nigeria ($1.1 \times 10^{12}$), Iran ($1.0 \times 10^{12}$), Venezuela ($0.74 \times 10^{12}$), Libya ($0.82 \times 10^{12}$), the Persian Gulf countries, Indonesia and some countries with a very limited domestic gas demand, such as Trinidad ($0.14 \times 10^{12}$), Ecuador ($0.17 \times 10^{12}$) and Colombia ($0.13 \times 10^{12}$). Considerable export potential also exists in Australia, since the large natural gas fields in the centre of the country and on the North Western continental shelf are remote from centres of gas consumption.

The demand for gas is concentrated in the industrialised societies of North America, Western Europe, the Communist Block and Japan, roughly along the lines indicated in Table 3.2.

TABLE 3.1

Reserves and Production, 1971

| | Reserves $(10^9\ m^3)$ | Production $(10^9\ m^3)$ | Ratio[a] |
|---|---|---|---|
| Western Europe | 4 540 | 96·3 | 47·2 |
| France | 200 | 7·1 | 28·2 |
| UK | 1 130 | 16·4 | 68·9 |
| Italy | 170 | 12·4 | 13·7 |
| Holland | 2 350 | 43·3 | 54·6 |
| Norway | 280 | — | ∞ |
| Austria | 20 | 1·9 | 10·5 |
| W. Germany | 390 | 15·2 | 25·7 |
| North America | 9 170 | 72·6 | 12·6 |
| Canada | 1 540 | 71·0 | 21·7 |
| USA | 7 630 | 655·0 | 11·6 |
| Central and South America | 2 060 | 41·6 | 49·5 |
| Africa | 5 460 | 4·7 | 116·2 |
| Middle East | 9 730 | 24·6 | 395·5 |
| Far East | 1 970 | 17·5 | 112·6 |
| Eastern Europe | 16 120 | 251·3 | 64·1 |
| USSR | 15 500 | 212·0 | 73·1 |
| Rest | 620 | 39·3 | 15·8 |
| Total | 49 050 | 1 162·0 | 42·2 |

[a] Years until reserves exhausted if production at present level.
Source: Veldorado 1971, Esso, A. G.

TABLE 3.2

Estimated Natural Gas Demand

Billion $(10^9)$ $m^3$ per year

| | 1970 | 1975 | 1980 |
|---|---|---|---|
| United States and Canada | 642 | 775 | 745 |
| Western Europe | 78 | 144 | 204 |
| Communist Countries | 237 | 344 | 450 |
| Australasia | 10 | 26 | 45 |
| Rest | 63 | 74 | 126 |
| Total | 1 030 | 1 363 | 1 570 |

It follows that local reserves will be insufficient to sustain gas demand at the levels expected in some of these areas. On the basis of 1971 reserves, for example, there would be only sufficient gas in North America to meet 13 years' demand at the 1975 level.[4] Similarly, reserves in Western Europe would last only 29 years, whereas the Soviet Block has sufficient reserves for 47 years.

On the other hand, what on the basis of simple statistics would appear to be a definite exportable surplus, may at times be strictly fixed to certain rates of oil production—if the gas is associated—or may already be earmarked for re-injection, for petrochemical manufacture or for future industrial use. Thus while both Eastern and Western regions of Venezuela will at some time have substantial gas surpluses, the complexity of crude oil associated gas production, gas re-injection programmes and planned domestic use will limit the surplus available for continuous liquefaction and export to relatively small volumes.[16]

While no undue importance should be attached to such calculations, as they can be fundamentally upset by new discoveries such as the gas fields in Norwegian waters of the North Sea or the recent finds off Ecuador and Western Australia, the figures do demonstrate a fundamental imbalance for certain regions of the world and indicate that gas supplies could be assured over a much longer period if major natural gas movements from regions of surplus to areas of deficiency could be arranged.

### 3.3.2 Elements of an LNG Scheme

The obvious way of correcting the supply/demand imbalance is by means of pipeline construction from remote gas fields to centres of consumption. This has been done in the past, e.g. in the US, Russia and Australia, but is impracticable in such long distance gas movements as those from North Africa to the US East Coast, Indonesia to Japan, Alaska to Japan, etc., and gas liquefaction prior to transport by sea is the obvious and only possible solution.

Major LNG projects which would help to balance fundamental gas shortages will, therefore, have a number of basic characteristics, and their constituent parts will also be fairly uniform.

In order to even consider shipping major quantities of gas from a field to a gas distribution company it will be essential, for example, that:

— field reserves should be sufficient to sustain a project over a minimum of 15 but preferably for 25 to 30 years;
— local domestic or industrial gas demand must be low and must be unlikely to grow substantially over the expected life of the project;

— gas in the quantities used must have a very low commercial value, i.e. there should be little or no alternative means of disposal such as petrochemical plants or power generation;
— the political climate in the area should be reliable. While relations between the gas producing and receiving countries need not be cordial they should at least be correct and the probability of a revolutionary change of regime in either which could lead to cancellation of the contract should be low.

It is essential that these conditions be met, since the investment required, even in the simplest LNG schemes, is extremely high. The risk involved in making such major investments must consequently be minimised. Furthermore, once base load gas, rather than peak gas only, is imported the normal functioning of industry, commerce and even of households would be threatened by any interruption of supplies. Particularly now, during the initial stages of LNG development, alternative sources of liquefied gas around the world will be rare or unobtainable if for some reason or other a liquefaction plant ceases to operate.

Politically and economically one can therefore differentiate between the components of an LNG project which are located at or near the gas field to which the above considerations apply, and those which are not located at or near the gas field. The components at or near the gas field are:

— the gas production facilities, i.e. gas wells, field lines, measuring and control equipment, pressure reduction and initial purification facilities, and well servicing equipment;
— the pipelines from the gas field to the liquefaction plant, generally from inland or submarine fields to a deep water harbour suitable for ships in the 100–150 000 tons dwt class;
— the liquefaction plant which normally includes gas purification, since frequently gas is piped impure and sometimes carries slugs of liquids from the well to the LNG plant all of which must be removed;
— liquid gas storage of sufficient capacity to load ships without causing delay;
— harbour and loading facilities for LNG suitable for large tankers, either of the standard type, i.e. jetties carrying pipelines to a loading pier head, or bow mooring devices with risers for a submarine loading line.

The components which are not located at or near the gas field, and which are no longer subject to local jurisdiction in the country of origin of the gas, comprise:

— mainly and most importantly, the ships equipped with special tanks for the carriage of LNG at −161°C;

— unloading facilities at the point or points of destination of the liquefied gas, again either special piers or mooring buoys, unless in some instances existing harbour facilities can be used;
— storage tanks for LNG capable of receiving entire ships' cargoes without causing delay;
— regasification facilities for the liquid gas;
— connecting pipelines with pressure regulators, measuring equipment etc. to link LNG vaporisers with existing gas mains.

It is proposed to discuss each of these components in greater detail in subsequent chapters.

### 3.3.3 Marine Gas Fields and Off-shore Liquefaction

A proposal to resolve some of the problems due to the political instability of certain gas producing areas has recently been made in the shape of a project to produce and liquefy natural gas at sea.[15]

Since many of the world's gas fields are off-shore, and since off-shore gas requires additional investment in the form of pipelines to the mainland and is therefore somewhat less likely to be tapped for local use, it was suggested that a liquefaction plant should be mounted on a barge or ship. A second vessel would provide the necessary storage, and tankers for the liquefied gas would be loaded by ship-to-ship transfer.

The advantage of such a system would be the possibility to move the liquefaction plant and gas storage facility to a new location if present arrangements turned out to be unsatisfactory. While, of course, most continental shelf gas fields are claimed by the nearest coastal states one could visualise a contract which would divide the ownership of the gas field from that of the liquefaction vessel; the latter could be registered elsewhere and would only buy natural gas, under a contract, from the producer.

The viability of such a scheme depends on the additional cost of ship- or barge-mounted liquefaction equipment over and above that of a plant on firm land. Furthermore it would have to be acceptable to the seller of the gas. Problems of wave motion, storms and weather at sea in general would also have to be overcome; repairs, overhauls and modifications of the liquefaction equipment might be difficult while on site. But in spite of these objections the idea seems of interest and may some day materialise. A brief economic analysis of such a plant is included in Appendix D.

### 3.4 Economics of LNG for Peak and Base Loads

Although in the case of LNG the expense of liquefaction and re-gasification is additional to the cost of transporting gas from field to consumer there is no doubt that the advantages of LNG as a peak load supply can compensate in many situations for this extra cost. After all, existing piped supplies are in many instances liquefied during the summer and stored for winter peak load use, thus adding much the same cost elements to the price of base load gas.

LNG as a base load fuel, on the other hand, must be considered an emergency supply to be used when cheaper local gas sources are insufficient or have been exhausted. Nevertheless it seems to be accepted that gas as a premium fuel, principally for household use, can command much higher prices than it has in the past. And while a substantial increase in gas price could eliminate large industrial consumers, who would be forced back to liquid and solid fuels, the domestic consumer can apparently afford the extra cost which would result from the partial replacement of piped gas by LNG.[5]

TABLE 3.3

Stake-Out Economics of an LNG Project

(Prices in US cents per million Btu)

|  | Typical piped gas supply | | LNG project | |
|  | Continuous[a] | Seasonal[b] | Continuous[a] | Seasonal[b] |
| --- | --- | --- | --- | --- |
| Wellhead gas price | 20–30 | 20–30 | 10–20 | 10–20 |
| Liquefaction, etc. Transport (ship or pipe) Regasification | 20–30 | 80–100 | 75 | 150 |
| Local distribution | 5–10 | 50–70 | 5–10 | 50–70 |
| Typical selling price | 45–70 | 150–200 | 90–105 | 210–240 |

[a] Industrial sales with high load factor.
[b] Domestic and small commerical sales with low load factor.

A brief economic analysis, summarised in Table 3.3, shows that liquefaction, sea transport and regasification will add about 75 ¢ per million Btu to the cost of a gas, assuming a sea voyage of 3000 miles and shipment at the rate of 1000 million scf per day. If LNG were to contribute 50% of total gas consumption this would increase a typical domestic price from $2·00 to 2·20 per million Btu, i.e. about 10%, not too much of a price to pay for continuity of supply in the face of developing shortages and rising energy prices.

If supplied to meet winter seasonal loads the cost of LNG would be higher, a minimum figure of $1·50 per million Btu being calculated in Table 3.3. However, this is less than a typical winter gas

B

price resulting from normal pipeline gas demand and commodity charges, concepts which are discussed in Appendix B.

It follows that LNG projects involving overseas liquefaction and long distance transport of liquefied gas should be economically attractive at the present moment for most seasonal and for certain year-round gas supplies. If in addition ecological and pollution advantages are taken into consideration the number of such instances will clearly rise. Finally, any further rise in competitive energy prices due to depletion of local resources will result in further LNG projects materialising.

## 3.5 GAS TRANSPORT IN THE FORM OF METHANOL

While the liquefaction of natural gas and its shipment in the form of LNG are well established technologically and the economics of major LNG projects can be evaluated with reasonable accuracy, there remains some nagging doubt in the minds of many engineers and economists whether LNG is, in fact, the cheapest method of transporting natural gas energy over large distances. Alternative methods must obviously be examined very carefully in order to establish that gas liquefaction-refrigeration is genuinely superior.

Among alternatives to be considered there are basically two processes, manufacture of Fischer–Tropsch hydrocarbons from synthesis gas, which in turn is made from natural gas, and production of methanol by established chemical processes, also via a suitable synthesis gas. Of these two, the Fischer–Tropsch route is not now being considered, mainly because it leads to a multiplicity of liquid products which it would be difficult and uneconomic to re-convert into gaseous fuel. Manufacture of a fuel gas of combustion properties similar to those of natural gas from methanol is relatively easy, on the other hand, and considerable thought has been given to the methanol route over the last few years.[1,2,3,10]

The schemes envisaged and evaluated by a number of contractors and producers of natural gas at the moment all involve conversion of natural gas into a methanol feedstock—carbon monoxide and hydrogen in the ratio of about one to two—by means of partial oxidation or steam reforming; this step is followed by methanol synthesis over a catalyst at high temperature and pressure, although pressures for this reaction have been reduced from 250–350 atm, to 80–100 atm in recent years. A temperature range of 300 to 400°C, while not too high to shift the methanol equilibrium towards hydrogen and carbon monoxide, is sufficient to produce reasonable reaction rates. The product under these conditions is not pure methanol but contains several percent of other alchols and has been given the name of 'methyl fuel'. The liquid without further purification can now be shipped in ordinary crude oil carriers, construction cost of which is between one fifth and one third that of LNG ships. Similarly, storage tanks for methyl fuel require less than one half the investment for LNG tanks. Finally, loading and unloading facilities, pipelines and instrumentation for methanol or methyl fuel are all substantially cheaper than their LNG equivalents.

On the debit side methanol is toxic and special precautions are required if spillage occurs. Furthermore, both plants to convert natural gas into methanol, and to convert methanol into substitute natural gas, the latter by low temperature reforming over a high nickel catalyst, are substantially dearer than their LNG equivalents, the gas liquefaction and LNG evaporation facilities respectively. An important consideration in some instances is the fact that in the case of the methanol route the synthesis plant, the section of the scheme requiring the highest investment, is tied to the gas well and therefore the host country. LNG ships on the other hand, which constitute the major cost item of an LNG scheme, are free-roaming and can be switched from one LNG source to another if necessary. Politically, therefore, LNG has certain advantages over methanol.

TABLE 3.4

Sea Transport of Methanol and LNG

|  | Liquid density $(g/cm^3)$ | Heating value $(kcal/tonne)$ | $(kcal/m^3)$ |
| --- | --- | --- | --- |
| Methanol | 0·792 | $5·33 \times 10^6$ | $4·23 \times 10^6$ |
| Typical LNG | 0·456 | $13·0 \times 10^6$ | $5·93 \times 10^6$ |

An important consideration in regard to marine transport is liquid density and calorific value per unit weight and unit volume. As shown in Table 3.4 methanol scores on density, but LNG is a superior material to ship, owing to its higher heating value per ton and per cubic metre.[10]

As to conversion efficiency, it seems that LNG evaporation losses during transport of the fuel and processing losses during liquefaction, together of the order of 15%, are lower than the corresponding energy and fuel expenditure when using the methanol route, which are about 30%. However, evaporation losses necessarily depend on the time of storage of the LNG, the shipping distance and the degree of insulation of storage tanks and ships. Similarly, fuel consumed for processing is to some extent a function of plant design, and some savings can be made when using the methanol route if methanol rather than reconstituted natural gas is consumed at the delivery end. Methanol schemes will therefore seem more attractive than LNG in situations where at least part of the product can be disposed of as a chemical rather than just a clean sulphur-free fuel. Fuel applications in which methanol scores over revaporised LNG are relatively few in number; present outlets for methanol as a chemical are, of course, mostly catered for by existing methanol plants, which produce the alcohol from high priced hydrocarbons at very much higher cost. And while it might be possible to sweep the market with low cost imports and put local manufacturers out of business, this should not result in sufficient outlets for justifying production on a scale comparable with recent LNG projects.

New outlets for methanol as a fuel would be essential, and several come to mind. While methanol can be regasified and the gas so produced can be distributed for domestic and industrial use, it would be preferable to use the alcohol fuel as such. It has, in fact, been proposed to fire boilers in air pollution sensitive areas which are remote from a piped gas supply; similarly gas turbines and other prime movers can operate on methyl fuel. In internal combustion engines methanol also has a number of advantages over petroleum distillates. Finally, methanol as a reducing agent in its own right and as a source of hydrogen and carbon monoxide for metallurgical use suggests a number of interesting possibilities, e.g. use in blast furnaces, as an injectant for direct reduction of iron ore in fixed or fluidised beds, and the reduction of non-ferrous metal ores such as zinc, copper and nickel oxides.

The economics of methanol and LNG transport depend entirely on the cost of shipping, as demonstrated in Table 3.5. Shipping costs are a function of shipping distance, carrier size and speed, and in the table it has been assumed that future LNG carriers will have a volume of 200 000 $m^3$ and an average speed of 16 to 18 knots. Methanol will be shipped in VLCCs (very large crude oil carriers) of about 200 000 tons dwt which would travel at roughly the same speed.

The table shows that under these circumstances the LNG shipping cost will outweigh the additional manufacturing cost and process losses of methanol when shipping distances are greater than 7000 miles. Over a distance of 12 000 miles, as for example from the Persian Gulf to either coast of the United States, LNG seems the more expensive alternative, even if methanol is regasified rather than burnt as a sulphur-free fuel. For the more usual shorter transport routes, on the other hand, and particularly if there is no outlet for methyl fuel, the LNG route seems the cheaper of the two.

However, there appear to be a number of potential schemes in the right range of shipping distances, gas volumes to be moved, and

TABLE 3.5
Economics of Methanol versus LNG Shipment
Movement of $5.0 \times 10^6$ tons of Methane Equivalent

|  | LNG schemes | | Methanol schemes | |
|---|---|---|---|---|
|  | 12 000 miles | 6 000 miles | 12 000 miles | 6 000 miles |
| INVESTMENT, $10^6$$ | | | | |
| Process plant | 160 | 160 | 321 | 321 |
| Off-site facilities | 81 | 81 | 83 | 83 |
| Start-up, spares, royalties | 20 | 20 | 42 | 42 |
| Plant storage, mooring, loading | 70 | 70 | 15 | 15 |
| Working capital | 33 | 33 | 42 | 42 |
| Ships | 1 100[a] | 640[b] | 275[c] | 140[d] |
| Terminal storage, unloading | 70 | 70 | 15 | 15 |
| Regasification | 5 | 5 | 25 | 25 |
| Total | 1 539 | 1 079 | 818 | 683 |
| Total ex ships | 439 | 439 | 543 | 543 |
| GAS PRICE, $\cent/10^6$ BTU | | | | |
| Wellhead price | 10·0 | 10·0 | 10·0 | 10·0 |
| Process fuel and loss | 0·3 | 0·3 | 8·2 | 8·2 |
| Evaporation loss | 3·0 | 3·0 | — | — |
| Capital cost: ships (13%) | 53·6 | 26·8 | 13·4 | 6·8 |
| all other (20%) | 32·9 | 32·9 | 40·7 | 40·7 |
| Operating cost: manufacture | 4·8 | 4·8 | 11·0 | 11·0 |
| shipping | 15·0 | 7·5 | 10·0 | 5·0 |
| other | 2·0 | 2·0 | 8·0 | 8·0 |
| Total | 121·6 | 87·3 | 101·3 | 89·7 |

[a] Ten LNG carriers, 200 000 m³, average speed 16 knots.
[b] Five LNG carriers, 200 000 m³, average speed 18 knots.
[c] Eleven VLCCs, 200 000 tons, average speed 16 knots.
[d] Six VLCCs, 190 000 tons, average speed 16 knots.
All investments based on Soedjanto, P. and Schaffert, F. W., Transporting Gas—
LNG v. Methanol, Oil and Gas J., June 11, 1973.

feedstock prices at the wellhead, where no clear-cut decision can be made on purely economic grounds. In such instances a number of other considerations will also have to be taken into account. Important criteria for a decision in favour of either methanol or LNG can therefore be summarised as follows:

— LNG will be preferred if shipping distances are short, methanol if they are long;

— methanol will be considered if there are outlets for it as a chemical feedstock or low-sulphur fuel, i.e. if only part or none of the imported material will be gasified;

— methanol will become more attractive once new large-scale applications, e.g. in metallurgy or as a transport fuel, develop;
— methanol plant construction will be either by locally owned companies in producing areas, or will be confined to politically stable countries in friendly relation with the gas purchasers and organisers of the scheme.

In conclusion, it would seem that some thought will have to be given from now on to the methanol alternative to LNG. A number of project studies with reference to Libya, Saudi Arabia and Iran are, in fact, progressing and detailed investigations for the shipment of methanol made by Petromin in Saudi Arabia to Piedmont Natural Gas Co. in North Carolina are said to be under way, the first instance of such a long distance movement.

## REFERENCES

1. Anon (1973). *Methanol versus LNG*, Pet. Press Service, Feb.
2. Anon (1971). Scale up problems defeat methanol as LNG alternative, *Eur. Chem. News*, **19**(485), 32.
3. Anon (1973). Humphreys & Glasgow look at methanol route, *Processing Engineering*, p. 9, March.
4. Backer, W. K. (1966). *Economics of Peak Shaving*, AGA Transmission Conference, Dallas, Texas, May.
5. Beebe, B. W. (1968). *Natural Gas in the American Energy Picture*, Paper 1, LNG–1, Chicago.
6. Dukes, W. W. (1967). Practical peak shaving economics, *Gas J.*, **332**, 28 and (1968) **333**, 45.
7. Gram, A. (1968). *Developing, Building and Operating an LNG Utility System*, Paper 6, LNG–1, Chicago.
8. McKenna, J. T. (1971). Status of LNG peak shaving in the US, *Pipe Line Ind.*, **34**(5), 33–36.
9. Nassikas, J. N. (1970). *A Regulator's View of the Growing LNG Industry*, Session VII, Paper 5, LNG–2, Paris.
10. Royal, M. J. and Nimmo, N. M. (1973). Big methanol plants offer cheaper LNG alternative, *Oil Gas J.*, **71**(6), 52–55.
11. Smith, W. H. and Anderson, P. J. (1972). LNG imports for the US, *Pipeline Gas J.*, **199**(7), 90–95.
12. Symonds, E. (1968). *Future of Natural Gas in World Energy Picture*, Paper 3, LNG–1, Chicago.
13. Vaughan, H. E. (1964). How Transco uses LNG for peak load shaving, *Pipe line Ind.*, **21**(1), 51–3.
14. Vincent, R. (1972). Peak shaving techniques by liquefied natural gas, *Gaz d'Aujourd'hui*, **96**, 293–8.
15. Zuber, H. (1971). FLOFF: floating offshore gas liquefaction and storage unit for processing offshore natural gas, *Gas J.*, **348**, 137–153.
16. Caro Ruben, A. (1972). *Venezuelan LNG Project, A Progress Report*, Session III, Paper 3, LNG–3, Washington.

*Chapter 4*

# Gas Liquefaction Plants

## 4.1 INTRODUCTION

The refrigeration and liquefaction sections of an LNG project represent the biggest single items of investment in such a scheme. Since natural gas has to be cooled down to below $-161°C$, the atmospheric boiling point of methane, it is generally necessary either to compress the gas before the first expansion step or, if it is supplied to the plant under pressure, to recompress after initial expansion, in order to achieve further cooling. For complete liquefaction several cycles of expansion and recompression are invariably needed. In addition, the low temperature of the process prevents the use of mild steel as a material of construction because of its tendency to embrittlement. Depending on mechanical duty, at least 5% nickel steel, stainless steel or aluminium alloys are used for all metal parts in contact with the liquefied gas. Finally, the presence of corrosive or high-boiling impurities will interfere with the operation of compressors, heat exchangers, pumps and fractionation equipment, and purification of the gas to a very high standard is essential. The need for large compression machinery, high-pressure equipment, special materials of construction, complex heat exchangers and large tankage, thus, all contribute to the heavy investment in the liquefaction complex.

## 4.2 GAS CONDITIONING AND PURIFICATION

Natural gas supplied by pipeline to a liquefaction plant from fields situated within about 100 miles of the plant, which may be inland or offshore, is as a rule only purified at the well-head to a minimal extent. This means that water, acidic gases, i.e. carbon dioxide and hydrogen sulphide, higher hydrocarbons and other impurities may be present in the liquefaction feed as gases or liquids, depending on pipeline pressure and ambient temperature. It follows that in order

31

to obtain constant flow through the line, it must be swept at regular and fairly frequent intervals with scrapers or pigs to prevent blockage and irregularities due to two-phase flow.

The first stage of any gas conditioning plant will therefore consist of a trap and facilities for the collection of liquid products. Depending on ambient temperature, water content of the gas and pressure drop, glycol or methanol may be injected as an antifreeze, and if this is done glycol or methanol recovery from the aqueous layer in the gas/liquid separator will be required. This usually involves fractionation in a small atmospheric distillation plant.

After reduction of liquid water, glycol and heavier hydrocarbons by simple gas/liquid separation, the gas is cooled by heat exchange to a few degrees below freezing; at pipeline pressure this results in further condensation, and additional water and heavy hydrocarbons separate out in a knock-out drum.

The process which follows is generally referred to as gas sweetening and serves to remove both $H_2S$ and $CO_2$. Sweetening reagents can operate by two basically different mechanisms, i.e. a reversible chemical reaction may take place between the acid gas and the solvent, or alternatively the acid gas may merely dissolve in the absorber liquid, in preference to and generally at a faster rate than the other gas components.[5,21]

In the case of physical absorption, equilibrium concentrations of $H_2S$ and $CO_2$ in the liquid are strictly proportional to the partial pressures of the gases. Reactive solvents, on the other hand, have absorption equilibria independent of the partial pressure of the gas, the ultimate concentration in the liquid being achieved by saturation.

Most solvents used for the absorption of acidic gas components are non- or only partially selective. They will remove carbon dioxide, hydrogen sulphide and mercaptans roughly in proportion to their original concentration in the gas. However, attempts have been made and have met with some success to separate impurities in the course of extraction rather than after their regeneration from the solvent. On the other hand, in those instances where only one of the gases occurs at any concentration, a concentrated stream of the impurity can be obtained in a single extraction. Furthermore, since most sulphur compounds extracted from natural gas are subsequently oxidised to elemental sulphur (Claus process) the presence of small concentrations of carbon dioxide in the regenerated gas is permissible.[28]

Typical chemically reactive solvents include aqueous solutions of most alkanolamines such as monoethanolamine (MEA), diethanolamine (DEA), diglycolamine (DGA), di-isopropanolamine (Adip), triethanolamine (TEA) and anthraquinone disulphonic acid (Stretford solution). In all these extractions, with the exception of the Stretford process, acid gases are absorbed at near ambient temperature by the alkaline compound and are released by heating to near its boiling point. The Stretford solution, on the other hand, which also contains sodium vanadate, sodium carbonate and a trace of chelated iron, when blown with air, oxidises hydrogen sulphide to elemental sulphur, which can be removed by filtration.

A series of absorption solvents based on potassium carbonate act in similar fashion to the alkanolamines. However, in the Benfield, Vetrocoke and Catacarb processes carbon dioxide reacts with potassium carbonate to form bicarbonate, which decomposes at elevated temperatures. A similar reaction takes place with $H_2S$; various additives, frequently arsenates, accelerate $H_2S$ removal by forming thioarsenates which decompose into arsenites and elemental sulphur (Giammarco Vetrocoke process). Catacarb and Benfield additives assist the rate of gas absorption by accelerating hydration of $CO_2$ gas.

Physical absorbents for acidic gases include anhydrous propylene carbonate (Fluor solvent), N-methyl-2-pyrrolidone (Purisol), the dimethyl ether of polyethylene glycol (Trigol). In certain instances physical absorbents need not be heated but can be flashed at reduced pressure to release the absorbed acidic gases. Their main disadvantage, compared with chemical absorbents, is their tendency to remove higher hydrocarbons from the gas, which is particularly undesirable where sulphur is to be recovered from the acid gas in a Claus plant.

The main disadvantage of chemical absorption is the highly corrosive nature of both absorbents and, particularly, absorbent-acid compounds. In an attempt to find an acceptable compromise, hybrid processes have been developed such as the Sulfinol extraction process which uses a mixture of the physical solvent sulpholane and chemical absorbents of the alkanolamine type.

Choice of a sweetening process depends on (a) original pressure of the gas to be treated—physical adsorption, for example, is greatly assisted by high gas pressure: (b) the original concentration of the acidic gases—chemical adsorbents generally speaking have a higher absorption capacity than physical solvents: (c) the relative concentrations of the acidic impurities—some solvents are more selective than others: (d) the permissible residual concentration of the impurities—again some solvents produce a cleaner final gas than others: and (e) the presence or absence of COS and $CS_2$.

Typical operating conditions for a number of reactive solvents are listed in Table 4.1.

A number of other solvents, both chemically reactive and physically absorbent, have been proposed and used commercially to sweeten

TABLE 4.1

Gas Sweetening Processes

| Process | Solvent | Type | | Typical concentrations Initial | | Residual (ppm) | |
|---|---|---|---|---|---|---|---|
| | | Chem. | Phys. | $H_2S$ | $CO_2$ | $H_2S$ | $CO_2$ |
| Adip | Di-isopropylamine | X | | Bulk | — | 0·5 | — |
| Benfield | Activated $K_2CO_3$ | X | | $H_2S/CO_2>1$ | Bulk | 30 | 150 |
| Catacarb | Activated $K_2CO_3$ | X | | $H_2S/CO_2>1$ | Bulk | 30 | 150 |
| Econamine | Diglycolamine | X | | | 3–20 vol % | 0·5 | 1 000 |
| Fluor | Propylene carbonate | | X | Bulk | Bulk | 0·5 | 1 000 |
| Girbotol | MEA | X | | 1–3 vol % | | 0·5 | 1 000 |
| | DEA or TEA | | | | 3–10 vol % | 1–2 | 1 000 |
| Purisol | N-methylpyrrolidone | | X | Bulk | Bulk | 0·5 | 3 000 |
| Selexol | Dimethoxypolyglycol | | X | Bulk | Bulk | 0·5 | 3 000 |
| Stretford | Anthraquinonedisulphonic acid, $Na_2CO_3$, $As_2O_3$, $Na_2VO_4$, chel. Fe compound | X | | <0·5% | Any | 0·5 | — |
| SNPA | Modified DEA | X | | Bulk | Bulk | 0·5 | 1 000 |
| Sulfinol | Sulpholane, alkanol-amine | Hybrid | | Bulk | Bulk | 0·5 | 3 000 |
| Vetrocoke ($CO_2$) | $K_2CO_3$, $As_2O_3$ | X | | | Bulk | — | 1 000 |
| ($H_2S$) | $Na_2CO_3$, $As_2O_3$ | X | | >1·0% | — | 5 | — |

natural gases from various sources. However, the above mentioned processes account for the bulk of modern gas purification plants.

After removal of acidic impurities by means of a chemically reactive solvent the gases are generally saturated with water, and this has to be removed before liquefaction. Three possibilities exist for gas drying.[14] It can be done by simple refrigeration, sometimes in a turbo expander, which allows the recovery of expansion energy for subsequent recompression. Both higher hydrocarbons and water are condensed and separated in a knock-out vessel.

Alternatively the gas can be dehydrated by means of a concentrated glycol solution. Di, tri and tetraethylene glycol are suitable absorbents at gas temperatures from 15 to 65°C; gas and dry glycol flow countercurrently through a scrubber, and the saturated glycol is continuously regenerated by stripping at 200°C. Gas dewpoints as low as −70°C can be attained irrespective of gas pressure; higher hydrocarbons, however, remain unaffected and have to be removed separately.

Thirdly, water can be removed by contacting with solid desiccants such as silica gel, alumina or molecular sieves. Gas to be dried is passed through a packed adsorption vessel under pressure, and both water and higher hydrocarbons combine by adsorption reactions with the desiccant. The latter is regenerated from time to time by passing hot gas, generally at a lower pressure, through the bed. Water and higher hydrocarbons are removed from the adsorbent and can be separated from the regeneration gas by cooling and condensation. Dew points of −50 to −70°C (both water and hydrocarbon) can be achieved.

If only higher hydrocarbons are to be removed active charcoal is the most suitable adsorbent. The use of certain molecular sieves, on the other hand, permits selective separation of water, hydrocarbons, sulphur compounds and carbon dioxide. However, capacity of solid adsorbents is insufficient to remove impurities present in very large concentrations and these processes are normally used for final gas clean-up rather than overall purification.

Permissible concentrations of impurities in natural gas depend on the choice of subsequent liquefaction process and particularly on the susceptibility to fouling and blockage of heat exchangers and expansion engines used for refrigeration. Generally it is desirable that water content of the gas should be less than 1 ppm. Carbon dioxide concentrations should be in the range of 50 to 150 ppm. Hydrogen sulphide, as far as potential fouling is concerned, could probably be as high as 30 to 50 ppm; in fact other considerations such as odour, corrosion and toxicity restrict it to a maximum of 3 ppm or less. Higher hydrocarbon concentration is not restricted provided liquefaction facilities are designed to separate liquids from gases in the course of refrigeration. However, it may be less expensive to pre-purify the gas than to provide extensive separation facilities in the liquefaction section of the plant.

## 4.3 GAS LIQUEFACTION

After the removal of the bulk impurities, natural gas destined for LNG production has to be compressed, cooled and liquefied.

As described in greater detail in Appendix A, refrigeration is based on the conversion of internal energy of a fluid into external work, and the second law of thermodynamics imposes a limit on the efficiency with which such a conversion can be carried out. To operate between two temperatures, e.g. the boiling point of the fluid at compressor exit pressure, $T_B$, and ambient temperature, $T_A$, a fluid undergoing a closed cycle would absorb a minimum amount of mechanical energy

$$W = Q \times \frac{T_A - T_B}{T_B}$$

in order to transfer energy equivalent to $Q$ from temperature $T_B$ to $T_A$. The efficiency term $(T_A - T_B)/T_B$ defines the ideal (reversible) energy required for refrigeration. Needless to say, the efficiency of actual refrigeration and liquefaction plants falls seriously short of such values.

Conversion of the internal heat of refrigeration into mechanical energy is generally effected by one of two mechanisms: either the compressed gas is expanded through an orifice and its temperature lowered by the Joule–Thompson effect, or alternatively energy is recovered by extracting work from the expanding gas in an engine.

### 4.3.1 Liquefaction Cycles

In order to liquefy a low-boiling gas it has to be cooled below its dew point. The latter is the temperature at which condensation starts taking place at a given pressure and for the main components of natural gas Table 4.2 lists dew points for a number of gas pressures.

To cool a gas, heat energy has to be removed from the com-pressed gas either by means of cooling water, if this is possible, or by means of an evaporating refrigerant if the temperature of heat removal is lower than ambient. From Table 4.2 it will be gathered that among the gases listed only propane can be liquefied by heat exchange with cooling water at moderately high pressure. All lower

TABLE 4.2
Approximate Dew Points of Hydrocarbons and Nitrogen (°C)

| Pressure, bar | 1·0 | 3·4 | 6·85 | 17·1 | 34·2 |
|---|---|---|---|---|---|
| Methane | −159 | −144 | −133 | − 92 | − 71 |
| Ethane | −91 | − 63 | − 44 | + 15·5 | — |
| Propane | − 46 | − 12 | + 12 | — | — |
| Ethylene | −104 | − 80 | − 62 | − 8·3 | — |
| Propylene | −49 | − 18 | − 5·6 | — | — |
| Nitrogen | — | −183 | −174 | −148 | −133 |

boiling gases require refrigeration by means of a refrigerant prior to a final compression step which results in liquefaction.

Since sensible heats of gases are lower than their latent heats, it is the latter which are almost invariably employed to transfer the bulk of the refrigeration energy. Intermediate refrigerants, if used, will therefore boil at atmospheric or slightly lower pressures, at some intermediate temperature between the boiling point of the gas which is to be liquefied, and ambient temperature. A single refrigerant will, as a rule, permit cooling by 60–90°C, and if an even lower temperature is required more than one refrigerant may be needed to act as intermediaries in the overall transfer of heat from the cold gas to cooling water or air.

The use of a series, or cascade, of refrigerants to cool low-boiling gases such as methane down to their dew point has given rise to the name of this process. Conventional LNG cascades usually employ propane, ethylene and methane as intermediate refrigerants, along the lines of the diagram in Fig. 4.1. Other combinations such as ammonia/ethylene/methane or Freon 22/Freon 13/methane are possible alternatives.

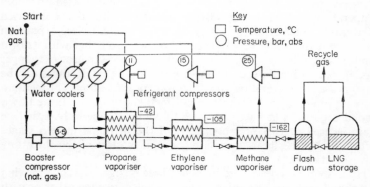

FIG. 4.1 Simplified flow diagram of a conventional refrigeration cascade.

The reasons for the use of a cascade rather than compression for cooling and expansion of a single component fluid are several. Firstly, as implied before, the pressure range over which compression would be required to reject heat at ambient temperature and absorb it at −160°C would be excessive. If methane were to be liquefied at 1·0 atm for instance it would have to be compressed at room temperature to about 1200 atm. Not only does this involve expensive equipment, but in addition the latent heat of vaporisation of all liquids tends towards zero near critical conditions; if use is to be made of the latent heat of the refrigerant, it is essential that the condensing temperature should be well below critical.

Similarly, evaporation of the refrigerant at very low pressure, in order to produce maximum cooling, requires a large vessel and a very big first stage compressor, both expensive items at the low temperature in question. The ranges over which refrigerants operate effectively are, therefore, both narrow and well defined.

An important consideration in the design of a liquefaction cascade is the temperature difference between the two fluids in each heat exchanger. The larger the difference the greater the irreversibility of the process and, therefore, the higher the energy consumption. On the other hand, the smaller the difference the larger the surface area required for heat exchange and, therefore, the first cost of the exchanger. Clearly a reasonable compromise must be found.

Operating conditions for a cascade process are therefore almost instantly defined once the number and type of refrigerants have been selected. Appropriate temperatures and pressures for a propane/ethylene/methane cascade, for instance, have been inserted into Fig. 4.1.

The diagram, although it shows the basic principles involved, is clearly oversimplified. Since cooling water or ambient air is the cheapest means of refrigeration each gas is first cooled to room temperature before being heat exchanged with the condensed refrigerant. Furthermore, each of the refrigerants, before expansion through the throttling valve, is cooled by heat exchange with its own vapour after release of the latter from the vaporiser heat exchanger (not shown). Finally, each of the higher boiling refrigerants can be expanded at more than one pressure, e.g. propane liquid at medium pressure pre-cools both ethylene and methane; this is followed by a second expansion at lower pressure to cool methane only. Ethylene again is used at two pressures to cool the methane in two stages before the latter's final expansion through the last of the throttling valves (not shown).

The system described here involves three refrigerants each moving in a closed cycle, i.e. not mixing at any stage with each other or with the liquefaction feed, which passes through separate passages in each heat exchanger and is thereby cooled gradually from ambient temperature to $-161 \cdot 6°C$, the boiling point of methane, which is, of course, the main component of most natural gases.

In addition to the standard cycle in which refrigerants exchange heat with each other and with the methane which is to be liquefied, there are other modified cascades.[1,12,20] While the refrigerants in a standard cascade are extraneous and each operates in a closed cycle, one form of modified cascade consists of a number of heat exchanges using the heavier hydrocarbons present in natural gas as intermediate refrigerants. In a modified cascade, therefore, refrigerant circulation is confined to a single stream, and by operating a number of stages of heat exchange at different pressures one can even eliminate the separate pumping of the individual heavier hydrocarbons. Furthermore, by using the heavier hydrocarbons, which are present as impurities in the natural gas, the outside purchase of refrigerants can be avoided. For instance, as shown in Fig. 4.2, a stream consisting of methane, ethane, propane and butane can be compressed to about 40 atm in a two-stage compressor, cooled with cooling water or by passage through air-fin coolers and split in a knock-out drum into a condensed propane/butane stream and an overhead consisting of the lighter components. Both streams, together with the natural gas feed, pass through the first multiple heat exchanger (E-1) at the inlet of which the heavy stream is expanded to compressor inlet pressure and returned through the third passage of the exchanger. The overhead from the rich separator, precooled in E-1, is split in a second knock-out vessel, the ethane separator, into condensate, which is expanded at the inlet of the second multiple exchanger (E-2) through which it passes in countercurrent with the light overhead, which also passes through E-2 and goes into E-3. After expansion to compressor inlet pressure it joins the low pressure propane/butane vapour from E-1 having first exchanged its cold with the liquid ethane and gaseous feed streams in exchanger E-2. Exchanger E-3 serves to further cool the natural gas feed and the uncondensed methane, which is expanded at its outlet to compressor inlet pressure; its flow is reversed and it is mixed with the other expanded gases after heat exchange in E-3. Liquefaction of the natural gas feed is effected by passing it as a separate medium pressure stream through the three heat exchangers in series and expanding the gas through a throttling valve after the last exchanger. Figure 4.2 shows diagrammatically and in very much simplified form directions of flow and heat exchange between the different streams.

Refrigeration in this type of gas liquefier is again due to heat exchange between the separate liquefaction feed and a series of refrigerants. However, as distinct from the conventional cascade, the refrigerants in the three heat exchangers are all derived from the same mixture of gases by fractional condensation at different temperatures. And while refrigerants and feed do not mix, i.e. each moves in a closed cycle, the refrigerant components all pass through the same compressors, i.e. they constantly recombine and are separated again in the knock-out drums.

It is also possible to design a modified cascade based on an open cycle, i.e. to use the heavier components of the feed gas to cool the feed by heat exchange or internal refrigeration. In other words, modified cascades can either be of the closed cycle variety, as shown

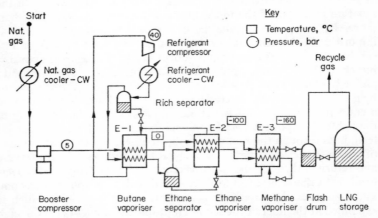

FIG. 4.2  Simplified flow diagram of a modified refrigeration cascade.

in Fig. 4.2, or employ an open cycle, as shown diagrammatically in Fig. 4.3.

The main advantages of a modified compared with a conventional cascade are the smaller number of compressors and heat exchangers required. Instrumentation, space requirement, and investment are claimed to be reduced and the system is very much simpler. Furthermore, there is no need for a supply of, and storage facilities for, intermediate refrigerants, and refrigerant losses are easily made up, e.g. by extraction of the heavier components of the feed gas. However, the close integration of heat exchange, compression and expansion leaves no room for flexibility, and final efficiency and performance of the plant depend entirely on the accuracy of the original design.

Advantages claimed for the conventional cascade process, on the other hand, are a lower overall compression horsepower requirement, rapid start-up and shut-down, since there is no need for preparation of refrigerants from the feed, and constant molecular weight of gases in each compressor. On the other hand, piping and compression are more complex, there are considerably more valves, and provision must be made for the supply and storage of the various intermediate refrigerants.

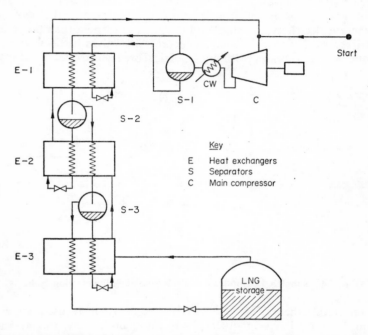

FIG. 4.3　Simplified flow diagram of a modified refrigeration cascade—open cycle. E, heat exchangers; S, separators; C, main compressor.

A hybrid type natural gas liquefaction plant in which the feed is first precooled in a propane refrigeration system, identical to that used in the first stage of a conventional cascade, and then passes into a multi-stage mixed refrigerant liquefaction system, is also widely used, and is claimed to be more efficient thermodynamically than the simple mixed refrigerant process, yet less complex in layout and operability than the standard conventional cascade. The propane precooled MCR process appears to be more widely used in modern plants than either of its competitors.

Very large conventional and modified cascade refrigeration systems for LNG have been built or are under construction. Single lines employing 135 000 hp of compressor power with gas compression of mixed refrigerants up to 450–500 psig and, in the case of conventional cascades, up to 200 psig (propane), 340 psig (ethylene) and 620 psig (methane) are in use.

A fundamental drawback of Joule–Thompson refrigeration cycles is their inability to recover and utilise mechanical energy released in the expansion of the gas. Any cycle which produces work by reversible gas expansion, which can be employed to compress the gas or to generate electricity, should therefore be more efficient than the series of irreversible expansions through throttling valves used in cascade liquefaction processes. Although practical experience does not always

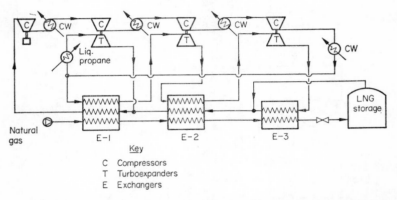

FIG. 4.4   Simplified natural gas expander—Claude liquefaction cycle.

support this theory, expander cycles are also widely used for gas liquefaction, alongside other plants using both conventional and modified cascades.

The expander liquefaction cycle is based on the compression of a mixed stream of fresh and recycle gas, generally in a two-stage machine, and passing the compressed gas in countercurrent with expanded gas through a heat exchanger. Part of the high pressure gas is withdrawn half-way through the exchanger and expanded isentropically in a reciprocating engine. The expanded cold gas which leaves the engine is used to cool the high pressure gas and is subsequently returned to the compressor. The remainder of the cooled high pressure gas is expanded through a throttling valve into a phase separator from which gaseous overhead is returned to the heat exchanger and the liquefied product is withdrawn.

Figure 4.4 is a diagrammatic representation of the Claude liquefaction cycle and gives some indication of both its advantages and its
drawbacks. It is essential, for instance, that the gas should not be
liquefied in the engine, and gas flow to the expander on the one hand
and to the throttling valve on the other must be carefully balanced
to achieve this. While it is easy to avoid liquefaction of single component gases the presence of heavier hydrocarbons, especially if their
concentrations tend to fluctuate, may upset this balance and result in
maloperation. It is common practice therefore to pass the gas
through a molecular sieve purifier before it enters the expansion
engine. Even so the expander cycle is simpler and the number of heat
exchangers, expansion valves and phase separation vessels is much
reduced compared with a conventional or modified cascade.

Modern LNG expander liquefaction plants use expansion turbines
rather than reciprocating engines and thereby claim to be able to pass
80% of the compressed gas through the expander and only 20%
through the throttling valve, thus obtaining almost complete liquefaction of the smaller stream. Nevertheless, in a recent comparison
of the two types[13] of liquefaction plant it has been claimed that well
designed cascades operate at the same if not a higher efficiency than
turbine expanders.

An alternative to the closed Carnot cycle in which heat is absorbed
at a low temperature by vaporising a liquid and is rejected by com-
pressing and liquefying the vapour at a higher temperature is the
so-called Stirling cycle. In this cycle a non-condensing gas is isothermally compressed at the higher temperature, cooled at constant
volume to the initial temperature, and recompressed, thus restarting
the cycle.

In order to achieve optimum performance the heat removed after
compression is stored in a regenerator and used to heat the expanded
gas before recompression. No external heating is thus required and,
as in the Carnot cycle, the heat energy is raised from the lower level,
$T_E$, to the higher level, $T_C$.

$$Q_E = W \times \frac{T_C - T_E}{T_E}$$

where $W$ is the mechanical energy supplied in compression and
released during expansion and $T_C$ and $T_E$ the temperatures of compression and expansion respectively. Again this applies in the ideal
case and actual machines using the Stirling cycle will be less efficient.

While in the original engine gas passes through the heat exchanger
at constant volume, in modern versions such as the Philips gas
refrigerating machine the cooling and reheating of the gas take place

at practically constant pressure, resulting in a refrigeration (or heating) cycle consisting of:

— isothermal compression;
— isobaric cooling;
— isothermal expansion;
— isobaric heating.

The mechanism through which this can be achieved is simple: two compression cylinders are connected through a regenerative heat exchanger and operate 90° out of phase. In the original design of a Stirling engine or refrigerator the two cylinders are in fact mounted 90° apart on the same crankshaft (see Fig. 4.5). In the later designs,

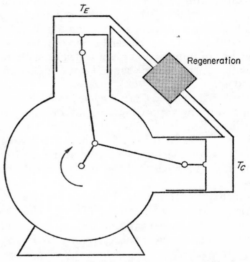

FIG. 4.5    Crank drive for Stirling compression cycle.

e.g. the Philips gas refrigerator, two pistons move in the same cylinder and the working fluid is transferred through ports situated between the two pistons which are again about 90° out of phase.

It will be appreciated that any Stirling engine can be operated as a refrigerator, a motor, or a heat pump. For instance, if the compression space is heated while the expansion space is at room temperature the machine will rotate and generate power; if heating stops and rotation continues the compression space will gradually return to room temperature while the expansion space will cool down. Finally, if rotation is reversed the expansion space will reach room temperature while the compression space will heat up eventually to bright red heat.

The multiplicity of commercial liquefaction plant designs is evident from Table 4.3 which lists a number of types advocated, amongst others, by the contractors named in column three of the table. Relatively minor variations in design have a significant effect on investment and plant efficiency.

TABLE 4.3
Commercial Gas Liquefaction Cycles and their Characteristics

| Type | Trade mark | Contractors[a] | Characteristic features |
|---|---|---|---|
| Conventional cascades | Several | e.g. Technip/Pritchard Phillips/Bechtel | 3 or more refrigerants, at least 3 pressures |
| Modified cascades | TEAL ARC | Technip/Air Liquide | Mixed refrigerant, composition natural, 2 pressures |
| | MRC | Brit. Oxygen/Air Reduction | Mixed refrigerant, composition adjustable 1 pressure |
| | PRICO | Pritchard | Mixed refrigerant, selected composition, 1 pressure |
| | MCR | Air Products & Chemicals | Mixed refrigerant, propane pre-cooled, 2 pressures |
| Turbo expanders | CPC | Airco/BOC | Nitrogen, closed cycle, several pressures |
| | | Rotoflow, Fluor | Open cycle |
| Stirling cycle | | Philips | Closed cycle |

[a] Most contractors with experience of cryogenic plant will design several different liquefaction cycles in accordance with local situations. A selection has been made of a number of designs used in the recent past, but the list is not by any means complete.

The difficulty of comparing different liquefaction plants and their cycles is due to the fact that neither initial nor final condition of the gas need be the same. Feed gas pressure can vary over a wide range and its temperature can be either above or below ambient. Furthermore only part of a high pressure pipeline supply may be liquefied and the remainder returned at a lower pressure. Heat sink temperature, i.e. cooling water or ambient air temperature, may vary over a range of 20°C approximately and there can also be variations in the operation of the plant, i.e. it can be a peak or base load plant. The liquid product may have to be merely liquefied and kept at atmospheric pressure, it may have to be maintained at low temperature for an extended period or it may be pumped away under pressure. Finally, feed gas composition may vary.

Instead of comparing different cycles it therefore makes more sense to express the efficiency of each plant in terms of a maximum, the latter being the minimum compression energy required to liquefy a natural gas under the conditions of the cycle, which is calculated in accordance with the methods discussed in Appendix A.

Calculated minimum energy requirements for a number of initial gas pressures and temperatures and three different heat sink temperatures are listed in Table 4.4. Only one gas composition has been considered, i.e. pure methane, and the product in each case is fully liquefied gas in equilibrium with its vapour at 1 bar external pressure.

TABLE 4.4

Minimum Work for Methane Liquefaction

| Initial gas pressure (atm) | Initial gas temp. (K) | Temp. of sink (K) | Minimum liquefaction energy | |
|---|---|---|---|---|
| | | | (kcal/kg) | (kWh/kg mol) |
| 1 | 300 | 300 | 261 | 4·86 |
| 1 | 314 | 314 | 284 | 5·28 |
| 15 | 300 | 300 | 162 | 3·56 |
| 30 | 300 | 300 | 137 | 3·02 |
| 50 | 300 | 300 | 118 | 2·60 |
| 35 | 289 | 289 | 119 | 2·62 |

Actual compression energy in existing and proposed liquefaction plants is substantially higher and liquefaction efficiency, i.e. theoretical divided by actual energy requirement, has been reported as 36 to 48% for various expander cycles. The efficiency of corresponding cascade plants, both conventional and modified, is in the range of 32 to 42%. It would appear, therefore, that the best expander cycles are slightly superior to the average cascade, while the worst compare unfavourably with the more efficient cascades. Stirling cycles have not so far been built in sufficient numbers to generalise about their performance.

The fact that all three types of refrigeration have been incorporated in a number of recently designed smaller liquefaction plants, as shown in the Appendix F, supports this conclusion. Larger plants, on the other hand, at present all use classical or incorporated cascades.

## 4.3.2 Compressors and Expansion Engines

The liquefaction of a gas, as discussed in the preceding section, is the result of heat removal by means of a refrigerant. The latter is generally a condensable gas and functions by absorbing heat at a lower temperature and rejecting it at a higher temperature; in order to do this the refrigerant has to be compressed and cooled at the higher temperature and expanded and heated at the lower temperature.

Compression equipment is therefore required for the heat rejection stage, and in cycles where expansion takes place against a balancing pressure and is used to convert heat into useful work (e.g. the Claude or Stirling cycle), expansion engines are employed as well. Compressors can be of a number of different designs and employ different principles. Table 4.5 lists the main types and sub-groups. Most compression equipment can also be used in reverse to convert a gas pressure differential into mechanical energy but the efficiency of some compressors if used as expansion engines is relatively low.[2,22]

TABLE 4.5
Compression Equipment

| | |
|---|---|
| Reciprocating compressors | Horizontal, single and multi-stage |
| | Vertical, one or more cylinders |
| | V-arrangement, multi-cylinder |
| | L-arrangement, 2-stage |
| Rotary positive displacement compressors | Screw type |
| | Vane type |
| | Lobe type (Roots blowers) |
| Turbo compressors | Centrifugal (radial) |
| | Axial (Barrel or split casing) |

When selecting a compressor or expander for a particular industrial process it is essential that the ratio of working fluid internal energy change to compression work should be high. Since working fluids or refrigerants are invariably gases, at least at some stage of the refrigeration cycle, the efficiency of compression must therefore be maximised.[6]

The type and conditions of compression in industrial equipment follow a path which is far from isothermal, i.e. exit temperature is invariably higher than inlet temperature; on the other hand it is not truly adiabatic since some temperature losses take place during compression, and a corresponding inward flow of heat occurs during expansion. Actual compressor operation is referred to as polytropic, and it is the polytropic compression efficiency which is of critical importance in compressor selection and compressor design.

In most industrial processes it is also important that plant operability should not be limited to a narrow range of throughputs and operating conditions; if, for example, compression efficiency falls off rapidly when the plant is running at a lower rate, or if the gas composition changes slightly when using certain types of compression equipment, whereas other types allow a more flexible operation, this would have to be borne in mind when selecting process equipment. The importance of running some types of compressors at constant speed is also significant.

Furthermore, in isothermal or near-isothermal compression and in refrigerant adiabatic expansion, partial liquefaction can take place, particularly if the refrigerant is made up of a number of gases. If part of the working fluid is present as a non-compressible liquid this may cause serious disruptions, and clearly some types of compression equipment can cope better with two-phase systems than others, either by handling the two-phase flow or by separating the two phases in the course of compression or expansion.

Since different types of compressors operate at different speeds and certain drivers, such as gas and steam turbines, are basically constant speed machines running at very high rotational velocities, the use of reduction gears is essential in certain instances, thus adding to the cost and reducing the inherent reliability of the complete installation. Diesel engines and similar reciprocating power sources, on the other hand, can be coupled directly with reciprocating compressors, in the shape of free piston engines which do not require any form of gearing.

A final consideration in the choice of a compressor is size and capacity. There are inherent limits in regard to maximum compression ratio, maximum pressure and maximum inlet volume for each type of compressor. These are functions of machine design, strength of materials and critical parts and also investment in the overall installation, i.e. compressor, driver and reduction gear.

The following brief discussion of the different types of compressors is slanted towards the above mentioned criteria, i.e. the advantage and limitations of each main group are listed and compared.

Reciprocating or piston machines can operate at very high gas pressures and also permit the highest pressure gain in a single compression stage, i.e. normally they operate at compression ratios of seven to twelve. Their characteristic feature is the presence of valves which open and shut during the compression process. This not only imposes limitations on throughput—the size of the valve in relation to the cylinder head and residual cylinder volume after compression —but also introduces an element of unreliability. Cylinder valves, whether mechanically or pneumatically operated, and particularly valve springs, have limited operating lives and need relatively frequent replacement.

Another source of potential trouble in reciprocating machines is the gas-tight seal between cylinder and piston. Piston rings or other sealing mechanisms are subject to wear and have to be replaced at regular intervals. Cylinder lubricants are also a potential source of gas contamination and nonlubricated compressors have to be used for the processing of ultra-pure gases.

The reciprocating motion of these machines and their drivers can be a source of intense vibration and the foundations for large compressors have to be remarkably solid; only relatively small reciprocating compressors can be mounted on movable frames.

On the credit side, piston machines are equally efficient over a wide range of operating conditions; throughput can be varied and running speed adjusted accordingly; compression ratio can be altered by a change in valve timing; many reciprocating compressors will cope with a limited amount of gas liquefaction; and finally extensive experience in the design and operation of these machines has resulted in equipment of extremely high reliability.

Little need be said about rotary or positive displacement compressors in the context of gas liquefaction. Most of these machines are of the blower type and the pressure gain which can be obtained is strictly limited to a maximum ratio of between 1·5 and 2. Furthermore, their construction does not lend itself to high pressure operation and both first cost and efficiency of most rotary compressors are low, although there are one or two helical compressors of the Lysholm type on the market that permit a somewhat higher pressure gain and also operate at a higher pressure. Rotary machines differ from turbo compressors in running speed; the latter can be coupled directly to high speed electric motors, gas and steam turbines, while positive displacement machines require reduction gearing. Not unlike reciprocating compressors positive displacement machines are also limited in size, i.e. inlet volumes of 10 million scfd and higher require two or more machines in parallel.

While reciprocating and positive displacement machines may be in use as auxiliary equipment in large gas liquefaction plants the principal compression equipment in major modern LNG and similar plants will be invariably turbo compressors, either centrifugal (radial) or axial. The two types differ in a number of respects.

In centrifugal compression the gas enters the casing, usually through a tangential duct, and is expelled by a series of shrouded impellers into a convoluted duct which returns it to the next impeller, where the gas is further compressed by centrifugal action, and so on. In axial compressors, on the other hand, gas flow, as implied by the name, is along the axis of the rotor. Blades, which are no longer shrouded, propel the gas along the length of the machine from one turbine wheel, by way of a ring of stator blades, into the next, gradually building up gas pressure. Efficiency and performance of the machine depend largely on the accuracy of these blades and any fouling, corrosion or mechanical damage will reduce both.[29]

It follows that liquid and solid impurities can be separated from the gas in the convolutions of the stator of a centrifugal machine, whereas they will pass straight through axial machines, and if they impinge on stator or rotor blades of one of the final stages of compression, considerable damage can result. Furthermore, it is relatively easy to introduce gas cooling between stages in centrifugal machines; in axial compressors, on the other hand, gas can only be cooled between two casings. In other words if interstage cooling is essential, axial compressors must be designed in the shape of several casings, rather than additional compression stages on the same shaft.

Axial compressors, on the other hand, have a number of advantages, particularly in the larger sizes. For the same performance they are distinctly smaller. Rotor diameter and therefore mechanical stresses on blade tips are reduced to about two thirds, which in turn allows larger capacities to be built without undue risk of material failure. The polytropic efficiency of axial machines is also claimed to be higher (0·87 versus 0·77) and although centrifugal machines are basically more flexible than axial compressors their lower efficiency at reduced throughput can be overcome by means of gas recycle from the discharge to the suction side and/or adjustable inlet vanes on the first stator ring.

There are relatively few limitations on maximum operating pressure; at the higher pressures casings of centrifugal machines are barrels; lower pressure radial machines and all axial machines are split horizontally for the insertion and exchange of rotors.

A feature of all turbo compressors is their tendency towards pulsating flow and pressure surges at reduced gas inlet volume. Throughput cannot therefore be regulated by throttling, and operating conditions must never approach the surge line on the compression ratio against inlet volume diagram. In axial machines the scope for inlet throttling is even less than in radial compressors and the provision of recycle facilities is that more important.

The availability of such features has in fact resulted in a basic design change of major compression equipment. While, until fairly recently, it was considered desirable to combine axial with radial machines, generally on the same shaft, to achieve maximum capacity and compression ratios, the preference, certainly in regard to major LNG facilities, has swung towards axial machines, and there is no longer any need for radial boosters. Both centrifugal and axial compressors operate at compression ratios between three and ten; the upper limit is imposed by the need for cooling between stages and also the difficulty of constructing excessively long rotors and casings. Maximum throughput of axial machines is as much a function of available driver size as compressor design and will be discussed after a brief review of available prime movers.

Compressors can be driven by internal combustion engines, electric motors, gas turbines and steam turbines.[8] It has been mentioned that certain drivers are especially suitable for particular types of compressors, e.g. diesel engines for piston compressors or electric motors for rotary positive displacement machines. The speed of electric motors is limited to a maximum of about 1500 rev/min, which is less than the requirement of turbo compressors, but higher than that of reciprocating machines—even if multipole electric motors are used to reduce driver speed. In either case reduction gears are needed. Furthermore, electricity is generally more expensive than other sources of energy, and start-up equipment is both expensive and prone to break-down. Electric motors rare arely made in sizes greater than 10 000 hp.

Major compressor installations are consequently driven not, as a rule, by electric motors but by gas or steam turbines. While some gas turbines are reasonably flexible and their efficiency does not vary greatly over a fairly wide speed range, this is not altogether so in the case of steam turbines. There are in fact two types of steam turbines commercially available, the variable speed machines of up to 55 000 hp, and larger drives, generally used for power generation, which are built in sizes up to 150 000 hp. Changes of rotational speed are achieved by adjustments of steam inlet vanes and exhaust steam flow on the smaller machines, while the larger units run at constant speed.

Two-shaft gas turbines of up to 50 000 hp are preferred as drivers for turbo compressors, not only because of their operability at varying speed and load conditions, but also because their output power is more easily matched to compressor start-up requirements than that of single-shaft machines which have a more limited range. The latter, being again mainly designed for power generation normally operate at constant speed and are available in very much larger sizes.

Gas and steam turbines are, as a rule, coupled directly to gas compressors to eliminate such problems as gear wear and locking of teeth as a result of the massive transmission of power from driver to compressor. Problems of alignment can be resolved by the use of flexible couplings. Typical engine speeds for gas turbines range from 3000 to almost 5000 rev/min. Most standard steam turbines appear to be designed for 4000 rev/min. The maximum rotational speed for turbo compressors (axial and radial) designed for flow rates of

50 000–250 000 ft³/min (24–120 m³/sec) at inlet condition falls into the range of 3000–5000 rev/min and thus eliminates the need for reduction gearing. Only relatively small centrifugal machines (up to 25 000 ft³/mm or 12 m³/sec) run at 8000 to 10 000 rev/min and therefore require a gear train.

Compressors in LNG plants are designed to perform a number of distinct duties. In conventional cascade plants propane, ethylene and methane are usually compressed separately, and typical compressor operating conditions for the three refrigerants and for a mixed refrigerant cycle are listed in Table 4.6, together with recommended materials of construction for the compressor casing.

Rotors are as a rule made of material meeting the worst operating conditions of the plant, i.e. for an inlet temperature of −160°C, one

TABLE 4.6
Compressors for LNG Plants

| Refrigerant | Discharge pressure gas (atm gauge) | Inlet temperature (°C) | Material of construction |
|---|---|---|---|
| Propane | 14–17 | −34 to −36 | Fine grain carbon steel (LCB) or 2% Ni (LC2) |
| Ethylene | 20–30 | −101 or −115 | 5% Ni steel or 4% Ni steel |
| Methane | 30–35 | −162 or −115 | 2DM Ni resist or 5% Ni steel |
| Mixed refrigerant | 25–40 | −34 to +40 | LCB or WCB |

generally uses Cr–Ni–Mo alloy steel. By doing this it should normally be possible to design compressors such that rotors in the different compressors are interchangeable and that the number of spare rotors held in stock can be reduced.

Since mixed refrigerant cycles not only require high discharge pressures but also involve a maximum flow of refrigerant through one or two compressors they present their designers with a problem. Clearly the cost of compression equipment can be lowered by increasing throughput to a maximum and reducing the number of trains in large plants to the operational minimum. In other words it would be desirable to process about 1000 tons of refrigerant per hour and to discharge the gas at about 40 atm, all in one machine.

Assuming operation at 3900 rev/min this leads to several conclusions; diameter and therefore rotor blade stress in a centrifugal machine would exceed acceptable standards and more than one compressor in parallel would be needed. An axial compressor, on the other hand, would be operating at a discharge pressure considered excessive by some authorities[27] for a horizontally split casing. Nevertheless at least one large LNG plant has been designed for this type of operation.[9,10] In the new LNG facility at Skikda, Algeria, there are three trains each consisting of low pressure and high pressure axial compressors with discharge pressures of 6·3 and 39 atm abs respectively. Refrigerant flow is 148 000 and 88 000 ft³/min and the two compressors are rated at 25 600 and 79 300 bhp. Both compressors are driven jointly by a single condensing steam turbine of a power output of 110 000 bhp. Only extended running of this plant can demonstrate the soundness of this concept, but there is no doubt that overall investment could be substantially reduced by this approach.

Although it is possible to convert the enthalpy of an expanding refrigerant into mechanical work by means of a reciprocating engine, in modern practice turbo machinery is invariably used. Both axial and radial machines can be employed and both types present certain problems in regard to condensation, mist formation and the impingement of droplets and erosion of metal parts by condensate.[13]

In the case of axial turbo expanders condensate, if formed during the initial stages of expansion, will tend to damage rotor blades near the turbine exit. Similarly, condensate formed in centrifugal machines can, owing to the high operating speed of turbo expanders, be centrifuged back into the rotor rather than collect in external catch ports.

Turbo expander efficiency is highest, much as in Joule–Thompson expansion, when operating at low temperatures, close to the dew point of the refrigerant. Mechanical efficiency of the machine will generally be at an optimum at rotational speeds between 10 000 and 50 000 rev/min, i.e. substantially higher than that of turbo compressors. It is not practicable, therefore, to couple the two types directly, and other forms of disposal of the energy generated in the machine may have to be found.

Where use of reducing gear or fluid drive to connect with a compressor is found to be uneconomic, turbo expanders may be used to generate electricity—again using gears to reduce revolutions to a more acceptable level—or if more convenient, the power may be wasted in an atmospheric air blower or turbulator.

The high speed of turbo expanders necessitates careful design; in particular, vibrations must be reduced to a minimum by careful rotor design. Rotor shafts tend to be of larger diameter than those of compressors, and pressure drop across each expansion stage must be controlled, a reasonable rule of thumb being an enthalpy drop of 50 Btu/lb of refrigerant (27 kcal/kg) per stage.

Temperature drop after start-up and temperature rise on shut-down must be allowed for in the design; frost formation and plugging of bearings and rotors must also be prevented by electric heating. Gas-sealed labyrinth bearings are common and seal gas pressure must be adjusted to result in a slight inwards leakage.

Capacity of commerical turbo expanders ranges from 50 to 5000 hp, and particularly the larger sizes permit the economical design of relatively small, i.e. peak load shaving, natural gas liquefaction plants.[31] LNG for export shipment, on the other hand, is not normally produced in expander plants.[30]

Clearly the design of compressors and expanders is a specialised field and will therefore not be covered here. Suffice it to say that the compressor must be matched, on the one hand, to its task of raising the pressure of a given volume of gas to the required pressure, on the other, to the characteristics of the gas in question. Particularly if the refrigerant consists of a mixture of gases, but even if it is made up of a single component, its pressure/volume relationship when compressed will differ from that predicted for an ideal gas. Compressor calculations will, therefore, have to take into account firstly the fact that real gases do not strictly obey the ideal gas laws and secondly that mixtures of gases have thermodynamic properties somewhat different from the simple average of those of their components. The design of compression machines and expanders for modified cascade refrigeration using mixed refrigerants will therefore be much more complex than that of simple compressors for conventional cascade liquefaction.[26]

### 4.3.3 Heat Exchangers for Cryogenic Systems

In both Joule–Thompson and expander cycles for gas liquefaction, though not necessarily in Stirling cycle refrigeration, it is essential to precool substantial volumes of gas to a level where further compression will result in liquefaction. Gas-to-gas heat exchange being notoriously inefficient, even if one of the gases condenses in the process, there arises a need for large heat transfer areas.[15,24]

The ordinary tube-and-shell exchanger is not suitable for cryogenic duties because of its steel construction, which is necessary to withstand gas liquefaction pressure, and its use in liquefaction plants is therefore limited to temperatures above −45°C. It is therefore used in gas liquefaction plants only to precool the gas by heat exchange with cooling water but not in the colder sections of the plant.

The original method of producing large transfer areas for cryogenic heat exchange were 'pipes-within-pipes' with gases flowing in opposite directions through central core and annulus (see Fig. 4.6). The advantage of this arrangement is the closer temperature approach between hot and cold gas which can be achieved by genuine counter-flow. The main drawbacks of concentric pipes are cost, high pressure drop and frequent blockages. Although still used extensively, pipes, fabricated in aluminium or copper and fitted with distance pieces to maintain the inner pipe in position, are rarely single strands wound on spools, as in the first gas separation plants, but are more often multiple strands connecting two or more manifolds, an arrangement which results in a lower pressure drop.

Although gas-to-gas heat exchange rates are in fact optimum under conditions of complete counterflow through long heat exchanger pipes, the high cost of the latter has resulted in a number of new approaches.

Instead of passing refrigerants in counterflow through concentric pipes it is easier, from a manufacturing point of view, to wind a skein of single pipes, all originating from a manifold, spirally on a

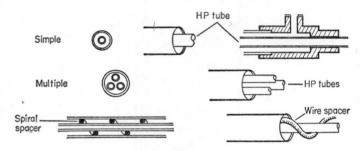

FIG. 4.6  Various forms of 'pipes-within-pipes'—heat exchanger surfaces.

mandrel and ending up in a second manifold, as shown in Fig. 4.7. Heat can be exchanged between the fluids in this first layer of pipes and those in a second layer, which is again spiralling around on the manifold, generally in the opposite direction, and between further layers of pipe and a refrigerant stream passing through the shell outside the tubes. Spirally wound exchangers of this type are very widely used in gas liquefaction plant; they are thermally efficient, fairly compact and robust and less prone to failing than are pipes-within-pipes. Very large areas, up to 18 000 m² in a single unit have been manufactured and incorporated in the larger LNG plants.

Heat exchange in gas liquefaction plants frequently takes place between an evaporating or condensing refrigerant and a gas. Clearly if evaporation or condensation is confined to one side only of the heat exchanger surface, the latter can have a smaller area than the surface in contact with the gas. The design of 'extended surface' exchangers is the obvious answer to the problem, and numerous versions of finned tube exchangers have been used in cryogenic and other heat exchange systems. Surface extensions can be genuine fins brazed or otherwise

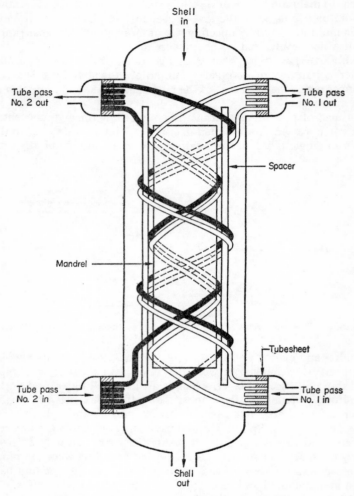

FIG. 4.7   Cross section of a spiral wound exchanger.

attached to the surface of the tube, or the surface of the tube can be porous with liquid entering into and partly flooding the pores of the metal. Other sections such as pins or rods can be used to extend the tube surface. Figure 4.8 shows an instance of a very compact finned tube gas-to-gas heat exchanger which has good structural rigidity and a relatively large heat transfer area, but is not suitable for high pressures.

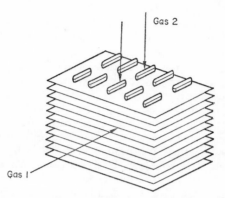

FIG. 4.8 Finned tube exchanger with flattened tubes and continuous fins.

An alternative design, the so-called plate-fin exchanger shown in Fig. 4.9 has recently been used in cryogenic, and particularly in natural gas liquefaction plants. Plate-fin exchangers have no tubes, but the gas passages are formed by sandwiches of metal sheet with alternating vertical and horizontal corrugations. The gases flow at right angles through the heat exchanger block, and although not ideal from a temperature equalisation point of view, the design provides a large heat transfer area at relatively little cost, internal soldering or brazing being limited to a small number of spots. This

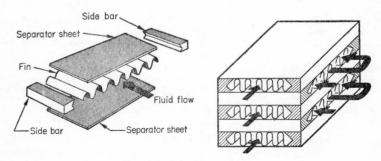

FIG. 4.9 Plate-fin exchanger.

C

type of core is relatively compact and suitable for low to medium pressure differences, but since it is usually manufactured by submerging the plates in a brazing melt and effective brazing requires uniform temperature distribution, maximum size of such exchangers is strictly limited (about 2000 m² per unit).

Construction of pipes-within-pipes exchangers is generally in aluminium or copper to provide flexibility and scope for coiling or reeling the pipes into manageable overall dimensions. Extended surface exchangers for low temperature work are, as a rule, made of aluminium; similarly the corrugated sheets for the construction of plate-fin exchangers are generally aluminium or one of its alloys.[15]

An important consideration in the selection of a heat exchanger is gas pressure; plate-fin exchangers are suitable only for relatively low pressure operation, up to 10 bar, or for situations where pressures are balanced. Higher pressures can be contained inside the tubes of finned tube exchangers, but not outside. For very high pressures, up to 140 bar, and large pressure differences, bundles of pipe-within-pipe exchangers and spirally wound exchangers are the most suitable.

The relative advantages and disadvantages of different types of heat exchangers in various cryogenic applications are summarised in Table 4.7; numbers in each column indicate the degree of preference for each type of exchanger.[11]

TABLE 4.7
Comparison of Cryogenic Heat Exchangers

| Type | Invest-ment[a] | Main-tenance | Fouling | Pressure | Com-pactness |
|---|---|---|---|---|---|
| Shell and tube[b] | 4 | 1 | 1 | 3 | 5 |
| Sintered tube[b] | 1 | 2 | 2 | 4 | 3 |
| Extended surface | 2 | 4 | 2 | 5 | 4 |
| Plate-fin | 3 | 4 | 4 | 2 | 1 |
| Tube-in-tube | 5 | 5 | 5 | 1 | 3 |
| Spirally wound (separate pipes) | 3 | 3 | 3 | 1 | 2 |

Order of preference—1 is first choice.
[a] Per unit heat transfer area.
[b] Temperatures down to −45°C.

Cryogenic heat exchangers have to be carefully insulated, and it is common practice to group them together in a 'cold box' the free space of which is filled with loose powdery or foamed insulating material. Valves located between exchangers have extended spindles and can be operated from outside. Alternatively, if the resultant package turns out to be too large and heavy, heat exchangers, e.g. for bulk LNG plants, are built as individual units, generally in the form of columns, which can be prefabricated in a workshop and erected on site and insulated.

Apart from heat exchange between cold and hot gases across a conducting metallic interface (recuperative heat exchange) gas refrigeration and liquefaction is often effected by alternately heating and cooling a heat store capable of retaining a substantial quantity of sensible heat (regenerative heat exchange). The principle of such

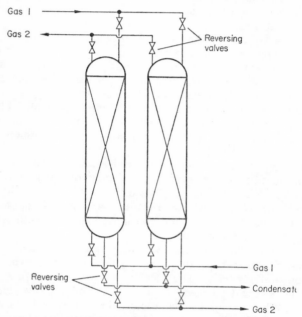

FIG. 4.10   Regenerative heat exchangers for cryogenic cooling of gases.

a refrigeration system with its reversing valves at the entrance and exit of the two regenerators is shown diagrammatically in Fig. 4.10. Again cryogenic regenerators are, as a rule, insulated in cold boxes with the spindles of the reversing valves, which should be made of low thermal conductivity material, extended to the outer container walls and operated from outside the box.[16]

### 4.3.4 Control Valves

All sections of a liquefied gas project require valves which should be carefully designed to match operational requirements. The latter are fairly standard in those sections which serve to supply and purify the gas, i.e. design and materials of construction must withstand gas pressure and chemical attack if any corrosive impurities are present. Since neither gas pressure nor hydrogen sulphide or other contaminants present unusual problems, only little need be said about the control valves required to operate the gas purification plant. An aspect to be borne in mind, however, is the high regeneration temperature which is sometimes required and the temperature shock to which the valve may be exposed. Globe valves with their relatively streamlined flow pattern, good closing characteristics, ease of replacement of worn or corroded seats and plugs are therefore preferred in this area.

In the liquefaction section of the plant certain other considerations apply. It is fairly easy to ensure tightness of stem seals and gaskets at higher temperatures when most packing materials have sufficient elasticity; at cryogenic temperatures, on the other hand, most natural and synthetic rubbers are altogether too brittle. Best results are obtained with ethylene fluoride polymers such as PTFE and Kel-F, but even here great care should be taken that no solids or ice are formed and forced past the soft stem packing which can be scored and cause leakage.

Similar considerations apply to metals used for valve body and trim construction. These must not lose their ductility at cryogenic temperatures, i.e. only austenitic stainless steels, copper, aluminium and nickel alloys are suitable. Furthermore, parts in contact with each other must have the same thermal expansion coefficient over the range from ambient to cryogenic.

Another important criterion in the selection of cryogenic valves is heat influx and refrigerant loss during cooling down, normal operation and servicing. The ideal valve will be leakproof, lightweight, its internal parts easily accessible and rapidly changed if necessary. While not all these characteristics can normally be attained in any one design the relative advantages and drawbacks of the different valve types have to be balanced against each other.

Butterfly valves, for example, are lightweight, their rotation shaft is easily sealed and they are cheap. However, they are unsuitable for large pressure drops and are generally used for throttling down rather than shut-off.

Ball valves also have a relatively low cool-down weight and no tendency to leak along the shaft. However, neither butterfly nor ball valve trim is easily accessible, and ball valves, if designed for access have bolted flanges in the insulated (cryogenic) area.

Globe valves for cryogenic duties are heavier, more expensive and subject to stem packing wear and leakage. However, they have recently been designed entirely without flanged connections in the area of contact of cryogenic fluids with the valve body. Their flow characteristics are also superior and correct design of valve trim will minimise all cavitation tendency.

Inspection and exchange of valve trim in modern cryogenic valves is facilitated by the provision of so-called warming extensions. The valve body, which is fully surrounded by insulation, is connected to external control equipment, not only by an extended stem, but also by a thin-walled stainless steel tube through which valve internals can be removed without disturbing the insulation. The extension tube is sealed sufficiently to prevent escape of liquid but is filled with vapour; it is usually installed at an angle to develop a vapour pocket, which acts both as insulation and seal against liquid leakage.[16] This is shown very clearly in Fig. 4.11, a partial section through a typical globe valve specifically designed for extremely low temperature operation.

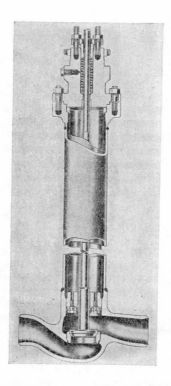

FIG. 4.11   Typical globe valve for low temperature operation.

# REFERENCES

1. American Gas Association (1968). *LNG Information Book*, Section II, New York.
2. Anon. (1972). Gas compression plant at Canvey Island, *Gas Oil Power* **68**(773), 59–61.
3. Anon. (1971). Current LNG technology, *Cryog. Ind. Gases*, **6**(5), 16–20, 22–24.
4. Anon. (1972). MCR process cycle offers base load benefits, *Cryog. Ind. Gases*, **7**(5), 37/8.
5. Beddome, J. M. (1969). *Current Gas Sweetening Practice*, Alberta Sulphur Research Quarterly Bull., Oct–Dec.
6. Brightwell, M. A. (1970). *The Selection of Compressors for Industry*, Paper 21, Inst. Mech. Eng. Conference on Compressors, Oct.
7. Brugerolle, J. R. (1970). *The Incorporated Cascade Cycle—Cold Production, Energy Generation*, Paper 2.33, 13th Int. Cong. of Refrig.
8. Bourget, J. M. (1968). *Selection of Refrigerating Compressor Driving Machines for Large LNG Installations*, Paper 29a, LNG–1, Chicago.
9. Bourget, J. M. (1971). Engineering large capacity LNG installations, *Oil Gas J.*, **69**(35), 71–80.
10. Bourget, J. M. (1972). Investment for large capacity LNG plants, *Oil Gas J.*, **70**(36), 74–75
11. Crawford, D. B. and Eschenbrenner, G. P. (1972). Heat transfer equipment for LNG, *Chem. Eng.*, **68**(9), 62–70.
12. Darradan, B. (1969). *Development of the Incorporated Cascade Cycle*, Proc. Int. Conf. LNG, pp. 240–266, London.
13. Dodge, B. F. (1972). Comparison of expander and cascade cycles for LNG, *Oil Gas J.*, **70**(24), 103–105.
14. Dow Chemical Co. (1962). *Gas Conditioning Fact Book*, New York.
15. Ducourant, D., Grange, L. and de Seyssel (1970). *Construction of Cryogenic Exchangers for the Liquefaction of Natural Gas*, Session II, Paper 7, LNG–2, Paris.
16. Goldfeder, L. B. (1972). Control valves for LNG facilities, *Pipeline Gas J.*, **199**(1), 58–74.
17. Harris, T. B. (1972). Natural gas treating with molecular sieves, *Pipeline Gas J.*, **199**(7), 76–89; **199**(10), 40–49.
18. Humbert-Basset, R. and Darrendeau, B. (1967). *Experimentation at the Nantes Station with the Air Liquide Incorporated Cascade Process for the Liquefaction of Natural Gas*, Paper 12, ATG Congress, Lyons.
19. Kinard, G. E. and Gaumer, L. S. (1973). Mixed refrigerant cascade cycles for LNG, *Chem. Eng. Prog.*, **69**(1), 56–61.
20. Linnett, D. T. and Smith, K. C. (1969). *Process Design and Optimisation of a Mixed Refrigerant Cascade Plant*, Proc. Int. Conf. LNG, pp. 267–287, London.
21. Maddox, R. N. (1970). *Conventional Sweetening Processes*, Proc. 49th NGPA Convention, pp. 98–103, New Orleans.
22. Murkett, D. G., Creswick, A. R. and Baker, J. W. (1970). *Review of Compressors and Expanders for Air Separation and Liquefaction Plants*, Paper 15, Inst. Mech. Eng. Conference on Compressors, London.

23. Naeve, K. L. (1968). *Operating Experience of a Small LNG Plant*, Paper 8, LNG–1, Chicago.
24. O'Neill, P. S., Gottzman, C. F. and Terbot, J. W. (1971). Heat exchangers for NGL, *Chem. Eng. Prog.* **67**(7), 80–82.
25. Pierot, M. (1968). *Operating Experiences of the Arzew Plant*, Paper 10b, LNG–1, Chicago.
26. Real, P. (1969). *The Influence of Real Gas Behaviour in Modern Compressor Design.* Proc. Int. Conf. LNG, pp. 288–299, London.
27. Schlatter, R. G. and Noel, C. J. (1972). *Axial Compressor Control in a Liquefaction Unit*, Paper 1/3, LNG–3, Washington; also (1973) *Oil Gas J.*, **71**(3), 53–55.
28. Stecher, P. G. (1972). *Hydrogen Sulphide Removal Processes*, Noyes Data Corp., Park Ridge, N.J.
29. Strub, R. A. (1970). *Design and Operating Experience on Industrial Axial and Radial Compressors*, Paper 6, Inst. Mech. Eng. Conference on Compressors, London.
30. Swearingen, J. S. (1968). *Design Considerations in LNG Expansion Liquefaction Cycles*, Paper 10a, LNG–1, Chicago.
31. Swearingen, J. S. (1970). Engineer's guide to turbo-expanders, *Hydrocarbon Proc.*, **49**(4), 97–100.
32. Thompson, C. E. and Sharp, H. R. (1969). *Kenai Plant Design*, Proc. Int. Conf. LNG, pp. 89–101, London.
33. Vogelhuber, W. W. and Parish, M. C. (1968). *Compact LNG System Using Large Stirling Cycle Cold Gas Refrigerator*, Paper 28, LNG–1, Chicago.

# Chapter 5

# Ships For LNG Transport

## 5.1 INTRODUCTION

In those instances where natural gas is liquefied for export from countries with a surplus to areas deficient in gas, the LNG is pumped from storage, through deep-water loading facilities, to ocean-going vessels suitable for the long distance transport of such a specialised cargo. LNG differs from most other bulk cargoes in a number of respects. Roughly by order of importance in regard to sea transport these are:

— its very low temperature (about −160°C);
— its low density (0·43 g/cm³);
— its high latent heat of evaporation (at −160°C this is about 535 kJ/kg);
— its low viscosity;
— its inflammable nature (5·3–14·0 vol % in air forms an explosive mixture).

The design of a carrier for such an unusual material is of necessity specialised and apart from the first ocean-going LNG ship, the *Methane Pioneer*, which was a converted crude tanker, all other LNG tankers have been designed from the start to carry LNG, the major alterations to cargo tanks, ballasting, cargo handling and safety features needed to adapt an existing vessel having generally been found uneconomic.

While LNG tankers thus differ from other ships mainly in the design and construction of their cargo tanks, there are other aspects of LNG sea transport, such as vapour recompression, cargo handling, use of regasified cargo as fuel and certain safety features, which also require discussion.[1,20,25]

## 5.2 LNG CARGO TANKS

The carriage of LNG, the bulk temperature of which is about −160°C, is complicated by the fact that mild steel when cooled

64

beyond $-50°C$ becomes brittle and disintegrates on impact. Apart from the necessity to insulate LNG to prevent evaporation loss it is therefore also essential to keep the LNG cargo away from contact with mild steel and particularly from the steel structure of the ship itself. Even a small leak in a cargo tank which would allow the liquid to escape and to make contact with the ship's plates or structural members could be disastrous.[2,25]

Metals which do not undergo major changes in physical properties and do not become brittle when cooled to $-160°C$ are, among others, aluminium, certain aluminium alloys, stainless steel, and 9% nickel steel. Membranes, structures, pipes, fittings, etc. in contact with LNG must be fabricated from such metals or alloys.

Furthermore, the thermal expansion and contraction of metals must be borne in mind. Any metal part periodically in contact with and cooled by LNG must be allowed to expand and contract freely; if this cannot be arranged for structural or other reasons the part must be thermally insulated in order to reduce the effect of temperature fluctuations, or must be made of alloys which have a minimal thermal expansion coefficient.

Low temperature insulation will clearly play an important part in the design of LNG tankage. While most forms of insulation, i.e. powders, foams or vacuum, can be used the low temperature makes it essential that water vapour, which would turn to ice and would upset the insulating properties of most forms of lagging, must be completely eliminated. Furthermore, powders will tend to pack solid at the bottom and leave voids at the top if the insulating space expands and contracts.

Very careful design of the cargo tanks and their surroundings is therefore essential and a number of basic approaches to the problem have been tried. Two basically different methods which have been used are the construction of self-supporting or free-standing LNG tanks, on the one hand, and the use of the hull, albeit carefully insulated throughout, as support for insulating layers and gas-impermeable membranes, on the other.

### 5.2.1 Self-supporting Tanks

A self-supporting or free-standing cargo tank in a ship can either be insulated by lagging attached to the tank with free space for inspection and detection of possible leaks available around the tank, or it can be keyed into position by means of insulation attached to both tank and ship's hull. In either case the tanks can expand and contract independently, i.e. they are not subject to thermal stresses. Also, being independent of the ship's hull, they are not exposed to quite the same risk if the ship is damaged by collision or impact. On the other hand, free-standing tanks do not fully utilise the available space in a ship's hold, in the light of the low density of LNG a serious waste of the ship's potential carrying capacity. The first of the transatlantic LNG tankers to use this principle, the *Methane Pioneer*, for example, was nominally a 5000 ton crude oil tanker, but after conversion was only capable of carrying 2000 tons of LNG in its five free-standing cargo tanks.[12]

The tanks made of aluminium were keyed into the original crude carrying space of the double bottomed tanker with balsa wood insulation on sides and floor. Tank roofs were insulated with fibreglass and the entire insulated space was vented with dry nitrogen to prevent ice formation. Sea water ballast was carried between inner and outer hull.[22,22]

Not surprisingly, the first grass-roots ocean going LNG tankers adhered to the free-standing cargo tank construction principle. The *Methane Princess* and *Methane Progress* had nine LNG cargo tanks, each subdivided into two compartments and one water ballast tank. Again the tanks were built of aluminium alloy sections, this time with angular fins to reduce cargo motion. Insulation consisted of plywood faced balsa, with a PVC foam seal between timber panels, and a fibreglass layer between the two layers of timber insulation, thus constituting a secondary barrier to LNG penetration.[3,4,5,6,32]

The first free-standing tanks made from metals other than aluminium were experimental tanks in the French ships *Beauvais* and *Pythagore*, the two vessels used to study and develop the stainless steel 'waffle' membrane technique. The internal gas-impermeable membrane, which was originally self-supporting but later developed into an integrated construction system, consists of an assembly of corrugated sheets. The sheets, which are lap-welded together, have wave shaped corrugations in two directions at right angles to each other. Materials of construction were stainless steel of Invar quality or at least 9% nickel steel.[7,16,27]

A further development in free-standing cargo tankage was produced by Esso in co-operation with Chicago Bridge.[22] Their tanks consist of a honeycomb structure of 9% nickel steel comprising an outer, insulation backed, and an inner membrane in contact with the LNG. Stiffeners connect the inner and outer walls. If failure of the internal membrane occurs LNG does not penetrate further than the secondary membrane and the structural strength of the sandwich remains unaffected. Manufacture of the tank and insulation can be contracted out and the entire structure can be delivered in one piece to the shipyard to be inserted as a single unit into the hull. While safe and comparatively reliable, this is one of the more expensive methods of tank construction.

An important feature of free-standing cargo tanks has been the need, at least in the past, to use plywood boxes and balsa wood load-carrying insulation. Modern developments in insulation such as polyurethane foam have changed this and self-supporting LNG tanks can be prefabricated and bedded into insulating foam produced *in situ*, rather than constructed completely in the yard or assembled from ready-made membrane insulation sandwiches.

A more recent free-standing design developed by Moss Verft in Norway on the basis of French patents involves the use of spherical vessels supported in their equatorial plane by a cylindrical skirt.[3,13,16,17] The vessels can be constructed in 9% Ni steel or aluminium alloys and withstand a minimum pressure of 0·7 atm gauge, i.e. LNG at slightly higher than atmospheric pressure can be contained. A secondary membrane consisting of non-permeable bonded polyurethane foam insulation reduces leakage, if it occurs, and ventilation of the ship's hold outside the tank removes LNG concentrations resulting from a 'small leak' (up to 300 mm$^2$). Potential savings in stainless steel consumption for spherical compared with prismatic tanks and integral construction are illustrated in Fig. 5.1.[30]

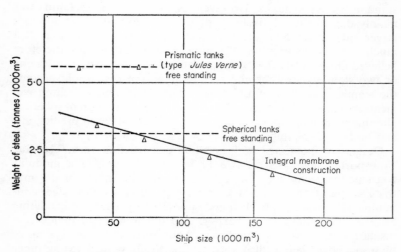

FIG. 5.1  Consumption of stainless steel in the construction of LNG tankers (after Guilhem and Richard).[18]

### 5.2.2 Integral Tank Construction

The advantages of integral construction of ship and LNG tank, along the lines of crude oil tanker design are:
— better utilisation of available space;
— less dead space to be monitored and purged;
— savings in alloy steel compared with free-standing tanks (see Fig. 5.1);
— same construction technique for all tanker sizes;
— no need for load-carrying insulation;
— rapid cooling down of tanks before loading.

Against these there should be mentioned the risk of LNG penetration to the inner and particularly to the external hull of the ship, the difficulties of welding the large membrane areas required, the impossibility of contracting out the construction of the tank, and the large thermal stresses involved in LNG tanks extending over the entire length and width of the vessel. In fact, integral LNG vessels frequently have a subdivided hold and a number of LNG tanks rather than a single container.

The first integral construction LNG tankers were built in Sweden, using the so-called Gaz Transport technique for Phillips/Marathon's Alaska to Japan LNG scheme,[14] and savings in stainless steel were found to be insignificant compared with the alternative of free-standing spherical tanks. However, as shown in Fig. 5.1, this was mainly due to the relatively small size of these ships—71 500 m³; larger integrated LNG carriers, which have been built since, have a much lower consumption of stainless steel and it was therefore believed fairly widely that in future integrated construction, with one or two stainless steel membranes confining the gas, would become the standard method of LNG carrier construction.[21,30] Improvements in free-standing tanks, particularly the construction of spherical aluminium tanks, has since cast some doubt on these predictions, and both types of ships are being built at the moment.

Designs of integrated tank and hull construction vary. The Gaz Transport system used in two Phillips ships, employs a modular construction technique. The interior membrane, made of 0·5 mm thick strips of Invar, a low thermal expansion stainless steel containing 36% nickel, is backed and supported by arrays of insulating boxes; a secondary Invar barrier, also 0·5 mm thick, divides the insulation into two separate layers, both inner and outer system of plywood boxes being filled with perlite. Studs welded to the inner hull of the ship support the entire structure.[11,18,19]

The alternative to the use of Invar stainless steel membranes, which do not become detached from the insulation because of their low thermal expansion, is the use of double corrugated 1·2 mm thick membranes, very similar in design to those used in free-standing cargo tanks in the Pythagore, but now attached to a continuous insulated lining of the inner hull. This type of construction under the name of Gaz Ocean/Technip is used on the French ship, *Descartes*, and while requiring somewhat more stainless steel than the membranes of the Phillips/Marathon ships, it does not need the high grade of steel of the latter.

Both types of construction originally involved the use of perlite filled plywood boxes for insulation. However, increasing labour costs have been eroding the cost advantage of perlite powder, which requires spacers and baffles to prevent settling and packing in the bottom of the ship. Polyurethane foamed *in situ*, it appears, is now largely replacing perlite, in spite of its higher raw material cost, since large areas of insulation can be formed using only minimal manpower.

While free-standing tanks can be fabricated, assembled, X-rayed, tested and installed in the hull in one piece, thus removing some of the more specialised construction techniques from the shipyard to the workshop, this is no longer possible with integrated construction. Although primary membrane/insulation/secondary membrane/secondary insulation sandwiches can be prefabricated, their size will be limited and extensive joining of plywood boxes and welding will have to be done in the yard.

Furthermore on the basis of the limited experience so far available integral construction LNG carriers may be slightly less reliable in performance than those with self-supporting tanks. At least the two ships on the Alaska/Japan run have experienced damage to one LNG tank each, whereas, so far, none of the free-standing tank carriers appears to have lost the use of a tank owing to gas escape. Careful weighing of the alleged greater reliability of independent tank construction in relation to their somewhat higher cost will be necessary when deciding on the method of construction of future vessels. No clear-cut advantage for either type can be deduced from the LNG tanker construction programme summarised in Appendix F, which includes similar numbers of both types.[15]

In fact technical developments in the sea transport of LNG are very much in progress. Proposals and experimental studies on the use of prestressed concrete, a material which will be discussed in connection with land-based tankage, have been made for the construction of ships' cargo tanks, and of entire barges for the transport of LNG. While not very ductile at any temperature reinforced concrete is not subject to embrittlement, and concrete ships' hull construction has progressed rapidly through the experimental stage. Integral reinforced concrete LNG carriers are a definite possibility.

The use of glass reinforced plastic (GRP) barriers within a cargo tank insulation system permits the elimination of at least one metallic barrier, and the replacement of the secondary membrane in the Swedish/French construction system used for the Alaska/Japan scheme by GRP seems perfectly feasible. Alternatively, the use of GRP-faced insulation on the *inside* of cargo tanks would permit tank or membrane construction from cheaper materials than high grade stainless steel; one proposal, in fact, suggested tank construction without the use of stainless steel altogether, a layer of GRP-faced insulation being backed by another layer of load-bearing insulation, again faced with GRP, and the complete sandwich being supported by the inner hull of the ship. A further proposal is to restrict the secondary barrier to the floor and lower third of the side wall of the tank or vessel, as it is unlikely that the entire insulation would ever be filled with liquid. While none of these ideas has so far been put into practice it may be significant that they have been discussed by representatives of Lloyd's Register, the world's largest ship classification society.[9]

Table 5.1 summarises containment systems developed by different designers and used in modern LNG carriers, and lists their chief characteristics.

**TABLE 5.1**
**Commercial Cargo Containment Systems**

| Designer | Self supporting | Pressurised | Characteristics |
|---|---|---|---|
| GAZ OCEAN | No | No | Single membrane with insulation serving as secondary barrier |
| CONCH | Yes (prismatic) | No | Secondary barrier forms part of insulation system |
| ESSO | Yes (prismatic) | No | Secondary metal shell structurally inter-connected with primary shell |
| TECHNIGAZ | Yes (spheres) | Yes | Externally insulated metal spheres supported by multiple rod and arm arrangement |
| KVAERNER-MOSS | Yes (spheres) | Yes | Aluminium spheres supported cylindrical skirt |
| GAZ TRANSPORT | No | No | Two Invar membranes separated by plywood insulation and backed by further lagging and inner ships' hull |
| IHI | No | No | Welded flat aluminium plate (15–25 mm) backed by two layers of insulation separated by a secondary barrier |

## 5.3 OTHER FEATURES OF LNG SHIPS

Apart from cargo tankage a ship for the ocean transport of LNG must have a number of other features which are more or less unique and connected with its task. In particular LNG carriers must have certain cargo handling and safety features, and in many instances the ship's engines will also be equipped to operate on vaporised LNG.

### 5.3.1 Cargo Handling Facilities

The LNG cargo is piped from the shore based tanks into the ship in low-temperature-resistant steel pipelines, which are carefully insulated to prevent excessive re-evaporation during LNG movement from shore to ship. Even so it is necessary to have recycle facilities to cool ships' tanks down and return the vaporised LNG to the shore for pressure maintenance in shore tanks, the inside pressure of which has been reduced by pumping LNG to the ship. Excess vapour formed is either reliquefied or vented to atmosphere.

During loading, the ship is linked to the pier or mooring buoy end of the pipeline by loading equipment which permits adjustments in height and distance to be made between the pier and the connections on board. The large diameter and weight of both LNG and vapour return pipes and insulation require heavy counterbalanced loading arms which can be accurately adjusted to meet their opposite flanges on board the ship. Loading facilities mounted on piers or jetties must be sufficiently flexible to allow for changes in tide and wind. Floating mooring buoys require less flexibility and movement but must withstand high winds and extremes of wave motion without capsizing.

Once aboard the ship the cargo is piped into the different cryogenic tanks, while simultaneously ballast water is drained from the separate ballast tanks so as to maintain the draft of the ship. It is common practice when unloading to leave a residue or heel of LNG in the tanks, which gradually evaporates during the return journey, thus ensuring that the tanks are precooled and that the next LNG cargo does not flash-vaporise when contacting the warm walls. In a new ship careful and gradual precooling of the tanks is necessary in order to protect the structure against thermal shock and to prevent the sudden generation of large gas volumes.

Problems have arisen in certain instances when the residual liquid in the tanks oscillated at a particular frequency and caused damage to containing membranes as a result. Careful calculation of the effect of pitch and roll frequency of the ship on cargo movement, particularly in near empty tanks, is essential to prevent shock waves developing at certain liquid levels in the tank. Alternatively unacceptable liquid levels may be avoided by pumping LNG from one cargo tank into another.

Access to the cargo tanks is invariably from the top, and both inlet pipes for liquid LNG and outlet pipes for liquid and vapour enter through central manholes set in trunks in the top of the tank. Similarly boil-off vent pipes equipped with low pressure safety valves are mounted centrally on each tank and communicate with a joint vent stack for the disposal of excess vapour.

Cargo is discharged or moved between tanks by means of cryogenic pumps, usually of the submerged variety. This eliminates the risk of vapour lock on the suction side when pumps are mounted on top of the tank, but presents such problems as electric connections in a potentially inflammable atmosphere, thermal contraction of casings, impellers, bearings etc. at the low temperature of the liquid. To avoid low temperature embrittlement, all parts of the pump and motor should be made of suitable materials, generally aluminium alloys.

A number of interesting points arise in connection with pump operation, removal and safety. The large size of the tanks and minimum number of overhauls means that the pump and motor should have the maximum possible service life. Also if a pump is withdrawn from the tank for maintenance there must be no leakage of air into or vapour out of the tank. Automatic non-return valves in the various pipes are essential and special fishing facilities must be provided if normally inaccessible pumps have to be exchanged or repaired. Pump motors, while always explosion proof, are also tested for operability in both gaseous and liquid methane. The electrical properties of liquid methane it so happens, do not interfere with the normal working of a squirrel cage motor, and in one design LNG is actually and intentionally pumped through the motor bearings and starter coils as a means of cooling. As a result classification societies and regulatory bodies now fully approve of the use of submerged motor driven pumps in LNG tankage.[9]

### 5.3.2 Reliquefaction Plant

While the principles employed for reliquefaction of LNG vapour on board a ship are the same as those of the full scale shore based natural gas liquefaction plant, there are a number of minor differences. Firstly, since the gas volumes to be liquefied are very much smaller, the use of turbocompressors or expanders, is no longer a foregone conclusion. In fact one standard plant for reliquefaction which can be used on board ship or to reliquefy vapour from LNG storage tanks consists of a closed cycle helium compression and expansion system with heat exchangers absorbing heat from the LNG vapour.

The Sulzer LNG reliquefaction cycle uses an oil-free two-stage piston compressor to raise helium gas to a pressure of 12 atm. The gas then passes through a heat exchanger where it is cooled by expanded helium gas returning from the expansion turbines. However, before entering the turbines the high pressure gas, now cooled to 33°C, flows through methane condensers mounted in the top of the LNG tank. It then returns to the two single-stage centripetal expansion turbines which reduce its pressure from 12 to 6 atm and cool the gas to about 21°C by converting the compression energy of the gas into heat, which is dissipated in an oil brake. Since the expansion turbines run at 85 000 rev/min power generation would require expensive gearing; the low efficiency of the cycle (12%) seemed a reasonable price to pay for the lower cost of such an expander.[28]

Other reliquefaction plants use modified cascades, invariably in a closed cycle to conserve refrigerant, or even conventional cascades. High liquefaction performance and plant efficiency are normally sacrificed in shipboard equipment in favour of low weight and plant volume. Reciprocating pumps are used, partly because of the smaller gas volumes to be compressed but also because of their flexibility; gas volumes for reliquefaction vary and can be much reduced by using LNG vapour as fuel for the main propulsion or auxiliary engines.

### 5.3.3 Utilisation of Vaporised Gas

While it is possible to reliquefy all the gas vaporised by heat leakage into the LNG tanks, and this is in fact essential when the ship is loaded with LNG and stationary, there are alternative means of vapour disposal. In particular, the ship's propulsion engines can be run on methane, whether they are of the steam turbine or diesel type, provided the boil-off gas is pressurised and pre-heated before injection.

Whenever LNG vapour is fired, provision must be made for dual firing; when the ship is running under ballast or if natural vapour formation in the LNG tanks is insufficient, there must be an alternative source of fuel, generally fuel oil or marine diesel, to meet the engine's fuel demand. On the other hand, the use of vaporised cargo as fuel is advantageous since it eliminates the need for reliquefaction and, by elimination of gas venting, reduces LNG loss. It is therefore customary to make provision for LNG vapour to be piped to some of the boilers or diesels either in addition to, or in place of, liquid petroleum fuels.

Since LNG boil-off would be insufficient to provide a ship's entire fuel demand it is more usual to fuel auxiliary boilers and engines with LNG vapour, rather than fuelling the main engine. Again this is satisfactory, provided dual fuelling has been allowed for, the cost of installing dual fuel burners and controls on such smaller boilers or engines being, of course, very much less than for the main propulsion engines.

The combustion of vaporised LNG in steam boilers or gas turbine combusters is relatively simple,[10] although in modern ships automation and computer control are used to ensure optimum combustion of the two fuels. One would not, as a rule, convert all burners to dual fuel firing and only provide sufficient burning capacity to cope with peak evaporation rate; in a well designed and properly insulated LNG ship this should be no more than 0·3% per day of the entire cargo.

In diesel propelled ships utilisation of methane vapour is less simple. Here provision must be made for mixing gas and engine air, i.e. the LNG vapour is aspirated rather than injected into the engine. This means that the engine will have to start up on its usual liquid fuel and that LNG will only be introduced gradually, once the engine is running.[29]

The same applies to auxiliary diesel engines operating rudder positioning mechanism, winches, hose handling equipment, electricity generators, etc. On the other hand, where steam engines or steam turbines are used as auxiliary power sources, no such problem arises.

### 5.3.4 Safety Aspects

Equipment on board a carrier to reliquefy vaporised LNG and to burn it constitute safety features in their own right. If for some reason LNG vaporisation occurs they provide an opportunity to channel the inflammable gas into a controlled process where it is either returned to safe storage or burnt. Under normal circumstances the two disposal methods ought to be capable of dealing with all the vapour generated in all the storage tanks.[28]

However, there may arise situations in which more gas is produced. If, for instance, there is a breakdown of the insulation or if LNG penetrates through the inner metallic barrier into the tank's insulation, vaporisation will be much more rapid and it may be impossible to consume all the vapour by the two mechanisms. Additional methane may then have to be discharged and this is usually done by venting to atmosphere, ensuring at the same time that the methane plume produced is kept well away from the ship's stack and hot propulsion machinery.

The propulsion machinery being located in the stern of the ship, gas must not be vented when the ship is steaming into the wind, but can be when there is a following or side wind. It is also desirable that there should be no other vessels downwind which the vapour plume could reach; similarly it would be impossible to vent LNG vapour when in harbour or near a populated coastline. Distances over which methane clouds can persist at different wind speeds and temperature conditions will be discussed in greater detail in Chapter 9. Suffice it to say at this stage that venting of vapour is an accepted practice and that it is safe except under the conditions mentioned. The same cannot be said of the discharge of liquid methane. Although methane gas is much lighter than air when at the same temperature and pressure, it must be borne in mind that freshly vaporised LNG will itself be at around $-160°C$; furthermore the air above an LNG spill will be cooled and the normal mixing process through outward diffusion will be slowed down. Finally, LNG of certain compositions can produce multiple explosions when spilt on water, a phenomenon which will be discussed in detail in Chapter 9.

The slow dispersal of spilt LNG, the occurrence of inflammable air/gas mixtures around the area, and the possible occurrence of small vapour explosions, make the discharge of liquid LNG a practice to be avoided as far as possible.

## 5.4 LNG SHIPS IN OPERATION AND BUILDING

A complete list of LNG tankers on existing LNG runs, ships under construction for future LNG schemes, and carriers built with a view to transport third party LNG cargoes will be found in Appendix G.

For a type of ship which only a few years ago made its first appearance on the high seas the growth in size of individual ships and their increase in numbers have been remarkable. The *Methane Pioneer*, subsequently rechristened *Aristotle*, was rebuilt in 1959 to carry 5000 m$^3$ of LNG across the Atlantic: today's LNG carriers on the Brunei–Japan run carry 70 000 m$^3$ ,and others now building are carrying 125 000 m$^3$ of LNG. Ships capable of carrying 165 000 and 200 000 m$^3$ are planned.

Including all smaller ships and prototypes built for test and development purposes a total of 16 LNG carriers were afloat at the beginning of 1973. By the second half of the 1970s their number will have been swollen to well over 100.[15]

## REFERENCES

1. Abrahamsen, E. (1963). Gas transport and ship classification, *Eur. Shipbuilding*, **12**, 24–39, 48.
2. American Gas Association (1968). *LNG Information Book*, Section V, New York.
3. Anon. (1961). Conch to ship liquid methane from Algeria to Britain, *Oil Gas J.*, **59**(22), 108.
4. Anon. (1963). Liquefied natural gas transportation, *Gas Coke*, **25**, 141–3.
5. Anon. (1963). Methane Princess, technical details, *Fluid Handling*, **163**, 263–6.

6. Anon. (1963). Methane progress launched, *Gas World*, **158**, 400.
7. Anon. (1965). New methane tanker design, *Oil Gas Int.*, **5**(3), 60.
8. Anon. (1972). Approval for new type of liquefied gas tank system, *IGE J.*, **12**(5), 148.
9. Atkinson, F. N. (1970). *A Classification Society's Approach to Vessels Designed for Carriage of Liquefied Gases*, Paper 1/7, LNG–2, Paris.
10. Balfour, T. T. (1970). *Gas Turbine Propulsion System for Large LNG Tanker*, Paper 11/4, LNG–2, Paris.
11. Caillaud, J. (1970). *The Use of Cryogenic Invar Alloy in the Sea Transport of Liquefied Natural Gas*, LNG–2, Paris.
12. Clark, L. J. (1960). *Sea Transport of Liquid Methane*, World Power Conference, Sect. Meeting Madrid, June.
13. Cranfield, J. (1972). Norwegian tankers employ new design features, *Pet. Petrochem. Int.*, **12**(6), 85–87.
14. Culbertson, U. L. and Horn, J. (1968). *Phillips–Marathon Alaska to Japan LNG Project*, Paper 13, LNG–1, Chicago.
15. Faridany, E. (1972). *Marine Operations and Market Prospects for LNG 1972–1990*, QER Special No. 12, the Economist Intelligence unit Ltd., London.
16. Gilles, M. A. (1967). *New Techniques for Ocean Transportation of LNG*, Int. Expo 67, Montreal.
17. Gilles, M. A. (1971). *Sea Transport of Methane*, Paper 8.15, 13th Int. Cong. of Refrigeration, Washington.
18. Guilhem, J. and Richard, L. L. (1970). *Lessons Obtained from Building and Starting Operation of 'Polar Alaska' and 'Arctic Tokyo'*, Paper 7/4, LNG–2, Paris.
19. Guilhem, J. (1971). *Construction of Large Methane Tanker—Why a Secondary Barrier*, 88 Cong. ATC, Evian, June.
20. Hunt, J. W. (1966). The techniques and rudimentary economics of transporting LNG by sea, *J. Inst. Petrol.*, **52**, 508.
21. Jackson, R. G. and Kotcharian, M. (1968). *Testing and Technology of Models of Integrated Tanks for LNG Carriers*, Paper 35, LNG–1, Chicago.
22. Latimer, D. M. (1968). *Esso Libya Venture*, Paper 15, LNG–1, Chicago.
23. Massac, G. (1972). Safety of the sea transportation of LNG, *Tanker Bulk Carrier*, **18**(10), 14–16.
24. Murphy, J. H. and Filstead, C. G. (1959). Ocean transport of liquid methane, *Gas Age*, **124**(8), 23–27, 30.
25. Nelson, W. L. (1954). Natural gas to move by barge, *Oil Gas J.*, **52**, 104/5, March 22.
26. Pastuhov, A. V. (1966). *The Transportation of LNG by Ship*, Adv. Cryogen. Eng., V–12, Proc. Cryogenic Eng. Conf., Univ. of Colorado.
27. Pilloy, M. and Richard, L. L. (1968). *Three Years' Experience with the Methane Ship 'Jules Verne'*, Paper 12, LNG–1, Chicago.
28. Ritter, C. L. (1962). Recent developments in liquefaction and transportation of natural gas, *Chem. Eng. Prog.*, **58**(11), 61–69.

29. Smit, J. H. and Steiger, H. A. (1970). *Slow Speed Dual Fuel Engines for LNG Tankers*, Paper 10/4, LNG–2, Paris.
— 30. Vrancken, P. L. L. (1972). Current LNG tanker designs, *Gas J.*, **349**, 173–7.
31. Ward, J. A. and Hildrew, R. H. (1968). *Importation of Liquefied Natural Gas from Algeria to the UK*, Paper 11, LNG–1, Chicago.
32. Zellerer, E. (1968). *Shell Structure Tanks for Transport of LNG by Ship*, Paper 37, LNG–1, Chicago.

*Chapter 6*

# Storage of LNG

## 6.1. INTRODUCTION

Storage facilities for LNG are required whether the liquid is to be used to meet winter shortages of gas or to supply base load gas by long distance shipment. In the latter case complete ships' cargoes have to be loaded into and unloaded from LNG tankers, i.e. storage capacity must be at least equal to the maximum volume of LNG expected in any one shipment. Storage for peak shaving, on the other hand, depends on the number of days per year during which gas is to be liquefied—200 to 220 in a temperate climate—and on the daily capacity of the liquefaction plant.

The vessels or space used for storage will constantly lose refrigeration, i.e. there will be heat leakage inwards and its extent will depend on the insulating quality of the containing walls. Rate of vaporisation or boil-off is a measure of heat leakage, and in order to design reliquefaction facilities correctly it is essential to predict the heat gain of the liquid. While this is relatively easy from a knowledge of the heat transfer properties of a man-made system, it will be much harder and the results will be less accurate where LNG is stored in the ground or in caverns, surrounded by frozen soil or rock of varying thermal and mechanical properties. Vaporisation losses from above-ground tanks are, consequently, much more predictable than those from underground or in-ground storage.[29]

## 6.2 TYPES OF LNG STORAGE

LNG on shore can be contained in double skinned metal tanks not dissimilar to those used in ships, i.e. aluminium or nickel steel inner vessels or membranes, surrounded by insulation and external weather-proofing. In addition prestressed concrete tanks can also be erected above ground, or can be cast below the surface. Finally existing caverns or underground spaces specially prepared for LNG storage can be used.[21] The main advantage of in-ground tanks, both concrete and natural, is that they do not require containment dykes to collect product from leaking or burst containers. The attraction of above-ground tankage on the other hand, is improved control of heat leakage and also the possibility of repairs.

## 6.2.1 Metal Tanks

The previously mentioned embrittlement of mild steel at temperatures below $-50°C$ makes it necessary to provide aluminium or stainless or at least 5% nickel steel membranes to contain the liquid and prevent it from making contact with external metal walls. A layer of insulation backs the gas-impermeable membrane; below the tank bottom insulation has to be of the load-carrying variety; powdered insulation is usually sufficient for the sides; glass fibre or a similar lightweight insulating material should cover the roof. The outcome of this method of construction is a container very similar to a ship's membrane tank which, instead of external support by the inner hull of the ship, is now held up by the external steel tank.[21,25]

Insulation between the two vessels must be sufficient to prevent cooling of the outer walls, and it is essential that the insulating space should be free of moisture to prevent ice build-up and loss of insulating power. It is also important to prevent extremes of temperature affecting the foundations.[14] Freezing and subsequent melting of the soil below the tank must be prevented by a combination of lagging of the tank and electric heating of the soil.

Foundations for storage tanks depend on soil conditions and tank size. For the larger tanks the main alternatives are either piling or a concrete ring beam, the former used where soils are soft, the latter, which is cheaper, where ground rock is not too far and the soil can be heated to prevent frost heave inside the ring.

The inner membrane of the tank can be produced in a number of alternative fashions. Since powder insulation in a double skinned vessel is subject to thermal movement and consequent compacting it is important that any motion of the inner skin be taken up in some way. In the case of a smooth membrane this is done by a fibreglass blanket which is sufficiently elastic to allow for inner tank expansion and contraction without transmitting any pressures to the outer powder insulation.[14] Alternatively the inner skin can be made of corrugated aluminium, which can be braced against the outer wall by ties of insulating material. Instead of working on the insulation, such a wall will absorb thermal expansion and contraction by flexure of the corrugations.

The third and most expensive means of stabilising the inner skin and protecting the insulation layer, is to use Invar, the 36% nickel steel which suffers practically no thermal expansion over the range of $-160$ to $+30°C$.

Construction of the tank roof is possible in a number of ways.

Since LNG temperature is close to the boiling point floating roof tanks are clearly out of the question. Similarly, freely vented fixed roof tanks cannot be used since explosive mixtures of air and gas could easily be formed. It is therefore essential to ensure that LNG tanks withstand a modicum of pressure or vacuum, and pressure roof tanks for LNG must be designed to contain atmospheric pressure fluctuations, the withdrawal of boil-off gas from under the roof by reliquefaction compressors and the vapour pressure fluctuations which will occur owing to changes in LNG temperature, e.g. sub-cooling below its boiling point, or changes in composition, e.g. loss of light components by preferential vaporisation.

The roof of the inner tank is normally an aluminium skin stretched between structural aluminium members. It is suspended by a large number of hangers from the outer dome and capable of supporting a layer of expanded perlite or similar insulation. A reinforcing metal ring round the circumference of the roof is designed to resist compacting forces exerted by the powder insulation between the two skins.

An alternative method of construction which promises to reduce the cost of LNG double skin tankage is to replace the inner metal roof by a suspended layer of insulation which is slung from the underside of the outer tank roof. In such a structure gas is no longer confined to the inner tank but the outer vessel is used to contain the pressure of the gas.[16]

The weight of metal LNG tanks is insufficient to ensure stability and the tanks have to be bolted down to the foundations in order to prevent lifting and movement when they are empty, i.e. filled only with natural gas. In the case of tanks with complete membrane separation between the inner LNG container and the outer supporting vessel this means that the retaining bolts must be spaced around the circumference of the inner vessel—in the light of the unsatisfactory mechanical properties of most steels at low temperature this implies use of stainless steel of at least 5% nickel content and a very careful welding procedure. Filling the insulation space with methane gas, on the other hand, permits locating all retaining bolts on the outside of the vessel, where they are no longer exposed to the low temperature of the tank content. In other words since there is then no pressure difference between outer and inner vessels it is sufficient to bolt the former in position, the latter being held purely by the insulation and bracing ties between outer and inner containers.

Although most LNG storage tanks are specifically designed and built for cryogenic duties, it is practicable, under certain circumstances, to convert existing tanks for liquid petroleum products into LNG storage tanks.

In one such instance a 1·8 million gallon tank in the N.E. United States was converted into a 1·6 million gallon LNG tank by casting a concrete slab as a floor base which housed both thermocouple junctions and hot water pipes to prevent frost heave if a gas leak occurred in the inner membrane. A rigid polyurethane layer supported the inside tank, which was fabricated in 9% nickel steel, welded to conform to ASTM A553. The annular space between the inner tank and outer wall was filled with a 2-in thick fibreglass blanket, to take up expansion and contraction of the membrane, and the remaining space was packed with polyurethane foam formed *in situ*.

Conversion of such a tank took only four months, compared with a minimum construction time of nine months for a similar grass-roots facility, and also turned out to be much more acceptable to environmentalists.

### 6.2.2 Prestressed Concrete Vessels

The design of LNG tankage built in prestressed concrete was originally developed in the United States by the Institute of Gas Technology and by Texas Eastern Transmission Corporation respectively.[1] While both designs use gas-tight metal membranes to separate gas and concrete the insulation is placed externally to the concrete in the IGT and between the membrane and concrete structure in the Texas Eastern design.[9] Special types of concrete are used to prevent cracking by thermal stress, and it is essential to reinforce with high tension wire or rods under tension, i.e. prestressing or poststressing.[8,22]

Both above and below ground location is theoretically possible with concrete tanks, and the thickness of insulation used will depend on whether the tank is free-standing or sunk. In the latter case the insulating power of the soil can be used to replace part of the insulation layer around the tank. In-ground storage also has the advantage of eliminating bunding, the construction of retaining dykes to collect spilt product if mechanical failure of a tank should occur, and it also scores on amenity value. On the other hand, the heterogeneous nature of the soil makes it difficult, if not impossible, to optimise design of the structure and particularly of the insulation.

In practice the problems encountered with in-ground LNG storage, particularly in the US, have been formidable. Particularly very large tanks, and clearly the larger the volume/surface ratio the more efficient this form of storage becomes, have failed owing to crack formation in the rock and frozen soil at the wall surface; heat transfer has exceeded calculated values owing to ingress of water, particularly where tidal water movements occur. Boil-off losses have become excessive in a number of tanks and some large in-ground LNG tanks have been built and subsequently abandoned.[29]

Above-ground LNG tanks in concrete, on the other hand, have been successful in a number of instances.[24] Construction technique usually consists of erecting (vertically) prestressed concrete panels to form the side walls. These are then wound circumferentially with high tensile steel wire, around which further concrete is poured while the wire is under tension. The resultant structure is covered with a metallic, prestressed concrete or glass reinforced plastic domed roof and insulated either on the outside with powder (perlite) insulation filling the space between concrete and an outer metal skin, or on the inside with load-bearing insulation faced with a gas-impermeable flexible membrane in contact with the LNG.

A type of concrete LNG tank developed by Technip in co-operation with Kellogg, has a waffled inner stainless steel membrane and a post-tensioned outer concrete wall. The latter is of monolithic construction, which eliminates the possibility of radial movement between wall and floor and can be built on piles 1·5 to 2 ft above ground to prevent frost heave. Interior plastic membranes (Mylar or similar films can be used) are also effective in preventing penetration of LNG into the polyurethane foam or concrete.[19]

The use of reinforced concrete for cryogenic structures is justified since its tensile strength is actually higher at low temperatures; the reinforcing steel, in spite of an increase in brittleness and a lower notch strength and ductility, is in fact higher in tensile strength, and when protected from impact by a layer of concrete, is capable of withstanding all expected mechanical and thermal stresses. In experimental tanks concrete has also been exposed to direct contact with LNG, without the use of membranes, and has stood up well to the ensuing changes in temperature. Future tanks, it is believed, will therefore be unlined, with heat insulation outside the concrete.[7]

An important consideration in the construction of concrete LNG storage tanks, is the avoidance of moisture penetration into the insulation space between outer and inner shells and also, if a leak develops in the inner barrier, to prevent LNG penetration into the concrete with resultant ice formation and possible structural damage. Recently a mastic coating, to be applied to the interior surface of the concrete, has been developed which, it is claimed, prevents both gas and water vapour leakage. A further advantage claimed for the product is that it reduces or eliminates the propagation of cracks in the concrete.[3]

### 6.2.3 Frozen Ground and Cavern Storage

There seems to be no basic reason why LNG and other cryogenic liquids should not be stored in natural or artificial caves or covered pits with the liquefied gas contained by frozen soil. In actual fact not all attempts to confine LNG by means of ice or frozen earth or rock have been successful.[11,25]

The procedure for excavating cryogenic storage is seemingly easy. First the ground is solidified by forming a base plug and cylindrical wall by circulating refrigerated brine through freeze tubes. While the outer freeze tubes are retained during subsequent mechanical excavation and the frozen wall prevents water penetration from the sides, the plug tubes have to be withdrawn and the bottom of the pit refrozen when excavation is complete. The roof of the pit is supported by a concrete ring beam on concrete foundations around the lip of the pit. The roof itself is constructed in aluminium, with a self-supporting membrane faced insulation suspended on the inside and extending underneath the concrete support. Alternatively insulation can be on the outside, or the roof can be hung from a steel dome.[15] Problems with this type of storage are, amongst others:

— movement of the ring beam and its foundations due to delayed frost heave;

— movement of pipeline trestles and pipes due to delayed soil shrinkage around the tank;

— higher than expected boil-off rates due to irregular freezing of the vessel walls.

They are accentuated by certain types of soil, and particularly by heterogeneity of soil structure. Thus different strata containing widely different concentrations of water can result in water migration and consequent soil movement. Similarly, tidal water which enters the storage area periodically can interfere with frozen ground tankage and its stability.

Commissioning of frozen earth storage must therefore take place over an extended period; in one installation it took 70 days to cool the tank down to LNG temperature by means of spraying, followed by a period of 90 days during which the liquid level in the tank was raised.[25] Boil-off, which is mainly caused by heat in-leakage, is at first a function of spraying rate, then increases gradually as the LNG level rises and eventually falls off to its equilibrium value as the frozen soil wall thickness increases.

As in shipborne and other forms of LNG tanks, pipework, pumps, instruments etc. are introduced from the top of the tank. Submerged pump/driver combinations are suspended from the roof supports and both pumps and instruments have to be highly reliable so as to reduce overhaul frequency and consequent heat leakage to the absolute minimum.[18]

A form of construction which is intermediate between in-ground containers, i.e. frozen holes, and above-ground metal tanks, is the in-ground container built originally above ground but subsequently surrounded by a soil berm which slopes naturally away from the tank, or occasionally, is retained by a supporting wall. If this berm has a minimum thickness of 10 ft, the container is no longer considered an above-ground tank with corresponding safety standards, and does not require containment dykes and enlarged safety distances. Furthermore, this type of construction permits better insulation and much lower vaporisation losses to be attained and eliminates the danger of seepage due to earth movement, tidal water flow etc.[20] In fact some of the world's biggest and most recent LNG storage schemes appear to favour this type of construction,[10] although a recent major explosion during cleaning and repairs of such a tank has occurred and its causes were, at the time of writing, not at all understood.

The use of mined or natural caverns suitably lined to prevent LNG seepage into the ground is even less common than successful in-ground storage. While mined caverns can be located in appropriate areas, provided suitable geological strata are present and faulting is limited to cracks which can be filled, natural caverns will rarely occur in locations where there is demand for stored natural gas and which are also accessible to transportation capable of carrying imported LNG. One instance of such storage in Massachusetts failed because it proved impractical to seal the leaks in the cavern wall, and the number of successful LNG storage caverns must be very small.[1]

While it would be difficult to quote actual prices for LNG storage which would have more than ephemeral value one can compare the economics of the different forms of storage. It is, however, essential to allow for gas boil-off and the (capitalised) cost of recompression. Under these circumstances it appears that apart from cavern storage, costs of which can vary enormously, the double skinned concrete tank without liner is cheapest, followed closely by in-ground storage, provided boil-off can be kept below 0·1% of content per day. In the same size range (20 000 tons plus) above-ground double skinned concrete tanks including a gas-impermeable membrane, although they have much the lowest boil-off rates (0·03–0·04% per day), are the most expensive, their total cost being only exceeded by in-ground storage with a boil-off rate of 0·3% per day, i.e. relatively unfavourable soil conditions.[11,12,20]

# 6.3 INSULATION OF CRYOGENIC VESSELS

Heat transfer from the surroundings to a cryogenic fluid can take place through the triple mechanisms of conduction, convection and radiation. The ideal means of limiting heat transfer, therefore, is to surround the cryogenic liquid with a vacuum through which neither heat convection nor heat conduction can take place. If, in addition, radiant heat transfer is minimised by providing reflecting surfaces there results a so-called Dewar vessel in which LNG can be stored for extended periods.

The problems associated with Dewar flasks are several: vapour pressure inside the flask must be very low (less than 1 mm Hg) and to withstand external atmospheric and internal LNG pressure a double skinned container will have to be strongly built. Also most materials, including metals, have a certain permeability to gases and to maintain the vacuum the vessel would have to be pumped out at regular intervals. Large Dewar vessels would be mechanically unstable since there must be minimum bracing between the two skins in order to limit conductive heat transfer. Finally the cost of the inner low temperature resistant membrane would be high.[1]

In practice therefore low pressure or vacuum in the space between the walls is rarely used as such and it is common to use insulating materials of low conductivity to prevent heat transfer. Most of these materials operate in two ways. Firstly they must be non-conductors of heat so that, even if they touch both walls, little convective heat loss occurs. Secondly, they must have a cell structure to prevent the circulation of gas in the wall space and thus the loss of heat by convection. If possible, insulating materials should be structurally strong to permit their use as a support for the thin and expensive interior membrane.

A further feature of low temperature insulation is the need to exclude moisture. While high temperature insulation will dry out under normal operating conditions and thus gradually increase its ability to contain heat, low temperature insulating materials must be protected from moisture which would deposit, build up into frost or ice, and substantially weaken the heat barrier set up originally. The wall space of a cryogenic tank if not under vacuum must, therefore, always be filled with dry gas.

In practice either liquid nitrogen vapour, which is of course absolutely water free, or vaporised LNG is circulated through the insulation space. If the latter is used there is no need to separate the insulating layer from the vapour space of the tank and LNG vapour can be allowed to spread into the insulation. This has the additional advantage of equalising pressures between tank content and insulation and permits a much lighter construction of the interior membrane of the storage tank.

Four types of insulating materials are used extensively in cryogenic systems.[14] The cheapest and most common insulant is a powder prepared by firing certain minerals, e.g. silica, diatomite, magnesia or asbestos, which expand under the action of heat and of evaporating moisture; such materials are marketed under trade names such as Aerialite or Perlite. Loose fill insulation is used to pack the wall space between the stainless steel membrane and the outer cylindrical walls of cryogenic tanks. When using it precautions must be taken against expansion/contraction of the two vessels, which results in packing of the lower regions of the vessels and slippage and empty spaces around the top of the cylinder. A further disadvantage of Perlite insulation is the need to empty the entire space before alterations can be made, instruments installed or repairs effected. Powder insulation is also unsuitable for use under tank bottoms where it would be exposed to pressure and excessive consolidation.

TABLE 6.1

Apparent Conductivities of Evacuated Powders

|  | Particle size | Apparent density $(g/cm^3)$ | Apparent conductivity $(microwatts/cm \times °C)$ |
|---|---|---|---|
| Silica aerogel | 0·025 microns | 0·10 | 21 |
| Expanded perlite | < 80 mesh | 0·14 | 10 |
| Diatomaceous earth | 1–100 microns | 0·29 | 10 |
| Alumina | 50–100 mesh | 2·0 | 18 |

Temperatures 76 and 300 K; pressure $10^{-1}$ mm Hg; thickness of insulation 2·54 cm.
Source: McClintock, Cryogenics, Reinhold, New York, 1964.

   The conductivity of loose fill insulation with a dry gas filling the spaces between the powder particles is several times higher than that of vacuum jacketed (Dewar) vessels. A hybrid form of insulation which combines some of the properties of vacuum jackets, i.e. low conductive heat transfer, and other properties of powder insulation, e.g. better support for the gas-impermeable membrane, is the evacuated powder filled jacket, frequently used for smaller cryogenic vessels such as vehicle tanks. Table 6.1 lists particle size, density and conductivity of a number of evacuated powders.
   The use of solid blocks of insulating material such as expanded polystyrene, PVC, phenol formaldehyde, foamed concrete, glass, rubber, balsa wood, etc. is essential if mechanical forces in addition to heat transfer are to be contained. While self-supporting blocks of solid foam still have to be protected from water penetration by means of sealing materials such as bitumen, mastics, surface coatings, resins

and waxes they can be used on the outside of single skin vessels, and as floor support, particularly if protected by sheet metal or sheet PVC.[2,6]

A third class of insulating materials is fibrous in nature: glass wool, rock wool, slag wool and Kaowool when attached to a rigid container can be used to take up thermal expansion and contraction owing to their inherent elasticity. They are also used as external tank roof insulation. Both preformed blocks and blankets of fibrous insulation are useful in that alterations and instrumentation work are possible without removing the bulk of the insulation. Both can be drilled through, the alterations made and the insulation patched around it.

The most recently introduced type of insulation is polyurethane foam produced *in situ* by foaming a mixture of di-isocyanate, polyalcohol and water in the space which it is proposed to fill.[14] The reaction in a so-called 'one-shot' system proceeds as follows:

1. Di-isocyanate + polyalcohol $\rightarrow$ urethane

2. Di-isocyanate + water $\rightarrow$ prim. diamine + $CO_2$

3. Diamine + urethane $\rightarrow$ crosslinked polyurethane

and since $CO_2$ is formed the resultant solid polyurethane has a foam structure. Sometimes the addition of a fluorocarbon (e.g. $FCl_3C$) is required to assist foaming. In other instances a prepolymer (reaction 1) is formed and introduced with the remaining ingredients.

Careful control of foaming is essential, but routine insulation of shuttered walls or double skin containers is now well established. Polyurethane foams are non-absorbent since each cell is closed, and require only a modicum of waterproofing; they are also elastic and self-supporting, thus combining the advantages of the previous two types. Apart from their application in a confined, shuttered space, polyurethanes can also be sprayed onto open surfaces; however, this is difficult to do effectively in the open air and satisfactory polyurethane insulation has so far only been applied indoors by spraying.

Insulating materials, whatever their type, i.e. loose fill, preformed blocks, fibrous blankets or *in situ* foams, must meet certain basic requirements. In particular their chemical and physical properties should not change extensively over their operating range; excessive embrittlement at temperatures around $-160°C$ would thus be unacceptable, just as melting and loss of insulating capacity at room temperature would be highly undesirable.

While it is not absolutely necessary for the insulating material in an LNG plant to be fireproof, it is at least desirable that the insulant should retard flames and, particularly, that it should not collapse entirely during or after a fire. A residual carbon skeleton such as that left behind by burning cork, is useful, while fire resistance as provided by expanded perlite, magnesite or asbestos is ideal.

## 6.4 MEASUREMENT OF TANK CONTENT

Measurement and custody transfer of LNG are complicated by a number of its features.[9] While volume of the liquid, if determined accurately, provides some indication of its commerical value, this is insufficient unless linked with simultaneous data on density and/or calorific value of the liquid and the vapour in equilibrium with the latter.[6]

LNG being at all times close to its boiling point even the measurement of liquid volume can be difficult owing to ill-defined liquid surface level; in addition dimensional changes in metering equipment, particularly tapes, may have taken place owing to temperature change. Depth gauging in LNG tanks must, therefore, allow for tape contraction. In order to determine the real liquid content of an LNG tank it must, furthermore, be borne in mind that vessel calibration at room temperature will no longer be accurate at −162°C and that a correction must be applied. A further allowance must be made in the case of a ship's tanks for inclination, i.e. list and trim of the vessel.[15] Depth gauging in a cryogenic tank is often done by temperature measurement by means of platinum resistance thermometers in a tubular sensor, as in the Trans Sonics custody transfer level system.[30]

The density of the liquid can either be measured, again allowing for instrument changes due to its low temperature, or it can be calculated from an analysis of the liquid and from its temperature, and from the corrected volume and density the liquid mass can be obtained.

In addition, the remainder of the tank (i.e. the ullage space) will be filled with vapour and the mass of the latter must also be included in the total. To calculate the vapour mass the ullage volume, vapour pressure, vapour temperature and gas composition (this can be calculated from liquid composition and temperature) need be known. Gas density can then be calculated and from the latter and the vapour volume one can obtain the mass of the LNG vapour in the tank.

When transferring LNG from ship to shore tankage, or vice versa, it is necessary to connect the vapour spaces of the two tanks to prevent the build-up of pressure or vacuum in either vessel. The result of balancing pressures is the reverse transfer of a volume of vapour equal to that of the liquid pumped in the direction of transfer. Allowance must be made for this effect in all calculations.[23]

A schematic diagram of the type of measurements which must be made and of their interactions is shown in Fig. 6.1.[17]

The calibration of LNG tanks, i.e. establishing a correlation between depth of liquid and its volume, much as that of any other tank, is effected either by direct measurement and calculation, or by the so-called water meter method. In the latter case measured quantities of water are introduced and the level of water in the tank is measured. The method is usually more accurate than measuring tank dimensions since irregularities in tank shape and the inevitable sagging of the measuring tape have to be corrected for if tank volumes are calculated from tank dimensions and liquid level.

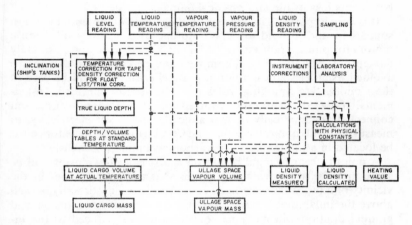

FIG. 6.1    Schematic guideline chart for static measurement of refrigerated hydrocarbon liquids.

In either case the low temperature of cryogenic liquids introduces additional problems. If tank volume is calibrated at 20°C the capacity of the tank at any other temperature, assuming linearity of expansion and the absence of constraints, is given by

$$V_t = V_{20}(1 - 3\theta)$$

where $\theta$ is the linear expansion coefficient of the structure between temperatures of 20°C and $t$. In fact, expansion over the range +20°C to −162°C is not usually linear and furthermore the expansion or contraction of a rigidly backed or supported tank does not follow the theoretical volumetric expansion rule. Calibration should, therefore, be carried to the lowest possible temperature and gauging correction tables must be prepared on this basis.[27]

D

## 6.5 BUNDING AND OTHER SAFETY ASPECTS

In order to guarantee against the danger of LNG spillage and over-flow with the attendant risk of explosions and fires it is essential to surround all LNG tankage with a retaining or bunding wall of earth or concrete in such a way that the entire content of the vessel will be collected within the walls if the tank fails when completely full, or alternatively to provide drainage into a bunded pond of the same volume. Under these circumstances the liquid will gradually vaporise and disperse, and provided no source of ignition is present there need be no further damage to property or casualties. On the other hand, if the spilt liquid is not contained and enters sewers, cable ducts and basements loss of life and physical damage can be tremendous.

It is very important, in consequence, to avoid all potential ignition sources in and around an LNG storage area. Clearly all electric motors for pumps, fans etc. must be intrinsically safe and absolutely flameproof. The use of steel hammers and percussion tools must be avoided. Neither spark ignition nor diesel engines should be used in the vicinity. 'No Smoking' rules must be strictly enforced. Instruments, if based on electricity, must be ventilated outwards and connected pneumatically or hydraulically, but never electrically, to measuring heads and valve actuators. Combustible gas detectors must be located in all potential hydrocarbon accumulation areas.

A point to bear in mind is that the worst risk of explosion will be some distance away from any ruptured tank or LNG line. In the vicinity of a failed tank methane concentration will normally be well above the inflammable range and only as a result of diffusion and gradual dilution will methane concentration be reduced to the inflammability limit.[11]

Spacing of tanks with respect to each other, with reference to the boundary of an LNG installation, and distances from buildings or equipment have been specified by the NFPA in their standard 59.A; Table 6.2 summarises their recommendations.

Similar safety distances are advocated by the American Petroleum Institute (API 2510 A), except that inter-tank distance is specified as one half the diameter of the smaller tank. In the UK the Institution of Gas Engineers is even more careful and recommends one half the sum of the two tank diameters.

In addition the IGE also recommend that each bunded area be provided with a sump where small spills of LNG would collect—rather than spread over the entire bunded area. All attempts at standardisation include specified distances from tankage to boundary of the property, to buildings and plant, particularly to fired furnaces and other possible sources of ignition. Safety distances can be

### TABLE 6.2
### Recommended Spacing for LNG Tanks (NFPA 59A)

|  | Above-ground[a] | In-ground |
|---|---|---|
| Distance between containers | ¼ sum of tank diameters | 10 ft min[b] |
| Container to associated equipment or buildings | 100 ft | 100 ft |
| Container to boundary or non-associated | 200 ft | the higher of 100ft or final frost line in soil |
| Minimum dyke capacity | 100% | None |
| Dyke to boundary | 100 ft | None |

[a] Provided an above-ground tank is subsequently buried and the liquid level is no higher than the backfill berm the rules pertaining to in-ground storage apply.
[b] This spacing may be insufficient for structural strength.

reduced—according to most standards if internal excess flow valves are fitted, but they have to be increased, according to other standards, where a sloping terrain enhances the likelihood of cold and therefore heavier gases diffusing along the ground. An unusual situation arises where a double-walled metal tank is surrounded by a pre-stressed concrete wall and berm, a practice adopted where land is expensive. Reduced spacing is considered justified under these circumstances.

In addition, thermal leakage, which could result in pressure build-up in the tank, must be counteracted by providing flaring and/or reliquefaction facilities[28] as well as pressure relief valves and alarms, all to be discussed in greater detail in Chapter 9.

## REFERENCES

1. American Gas Association (1968). *LNG Information Book*, Section III, New York.
2. Anon. (1971). Polystyrene foam: LNG tank insulation, *Cryog. Ind. Gases*, **6**(5), 39.
3. Anon. (1972). Concrete LNG storage tank uses mastic barrier, *Pipeline Gas J.*, **199**(12), 60/61.
4. Bodley, R. W. (1969). *Design Considerations for LNG Storage Tanks—Current US Practice*, Proc. Int. Conf. LNG, pp. 457–468, London.
5. Boyle, G. J. (1972). Basic data and conversion calculations for use in the measurement of refrigerated hydrocarbon liquids, *J. Inst. Petrol.*, **58**, 133–7.
6. Burkinshaw, L. D. (1968). *PPO Foam for Cryogenic Applications*, Paper 33, LNG–1, Chicago.
7. Clapp, M. B. and Litzinger, L. F. (1971). Design of marine terminals for LNG, LPG and ethylene, *Pipeline Gas J.*, **198**(7), 72–82.
8. Closner, J. J. (1970). *LNG Storage with Prestressed Concrete*, Paper 32, LNG–2, Paris.

9. Cofield, W. W. and Graves, R. G. (1971). Measuring LNG is important, *Gas*, **47**(3), 63.
10. Ewing, G. H. and Smith, E. L. (1971). Texas Eastern unveils Staten Island LNG plant, *Gas*, **47**(1), 28–31.
11. Gibson, G. H. and Walters, W. J. (1971). Consider safety, reliability, cost in selecting type of LNG storage, *Oil Gas J.*, **69**(1), 65–69.
12. Gibson, G. H. and Walters, W. J. (1972). Landbased LNG storage tanks, *Gas J.*, **349**, 179–84.
13. Hale, D. (1971). Underground storage, LNG plants expand for biggest supply ever, *Pipeline Gas J.*, **198**(14), 19–23.
14. Hall, R. C. (1968). *Thermal Insulation for Refrigerated Chemical Plant*, Review of Refrigeration Tech., Inst. Mech. Engs., Cleethorpes, March.
15. Hamilton, W. and Manning, G. P., *Design and Construction of Roofs over Inground Reservoirs*, Proc. Int. Conf. LNG, pp. 365–379, London.
16. Hatfield, J. and Attfield, R. A. (1969). *LNG Storage at Ambergate, Derbyshire, England*, Proc. Int. Conf. LNG, pp. 419–434, London.
17. Hunt, W. (1972). Special considerations related to large scale refrigerated LNG/LPG custody transfer measurements, *J. Inst. Petrol.*, **58**, 158–63.
18. Hylton, E. (1969). *The Use of Submerged Electric Motor Driven Pumps for Liquid Gases*, Proc. Int. Conf. LNG, pp. 379–393, London.
19. Kellogg, M. W. (1972). *Liquefied Natural Gas*, Bulletin, Dec.
20. Lanterbach, G. M. (1966). *Technical and Economic Considerations for the Selection of Surface and In-ground Tankage for LNG*, Natural Gas Symposium Proceedings, pp. 91–111, Institute of Petroleum, London.
21. Lusk, D. T. and Dorney, D. C. (1972). LNG Storage Tank Systems are Safe and Versatile, *Oil Gas J.*, **70**(41) 91–95.
22. Orcharenko, A. P., Morozor, N. S. and Chepiga, E. V. (1971). Ferroconcrete isothermal reservoirs, *Stroit. Turboprod.*, **16**, 9–11, June.
23. Thomeson, C. H. W. (1972). Flow measurement of refrigerated hydrocarbons, *J. Inst. Petrol.*, **58**, 138–42.
24. Turner, F. H. (1972). Prestressed concrete LNG storage tanks, *Gas J.*, **349**, 189–90.
25. Ward, J. A. and Egan, P. C. (1969). *Experience with Frozen Inground Units for LNG—Canvey Island, UK*, Proc. Int. Conf. LNG, pp. 342–364, London.
26. Wardale, J. K. S. (1969). *Storage of LNG in Metallic Containers*, Proc. Int. Conf. LNG, pp. 435–456, London.
27. Watson, P. B. (1972). Instrumentation for the measurement of refrigerated hydrocarbon liquids, *J. Inst. Petrol.*, **58**, 126–32.
28. Wicker, P. (1971). Reliquefaction of LNG boil-off gas, *Oil Gas J.*, **69**(3), 53–55.
29. Zellerer, W. (1969). *Problems in Connection with Tanks Containing a Lower Temperature Media*, Proc. Int. Conf. LNG, pp. 394–418, London.
30. Blanchard, R. (1972). *Shipbuild. Shipp. Rec.*, **120**(19), 21–22.

*Chapter 7*

# Revaporisation and Distribution of LNG

## 7.1 INTRODUCTION

LNG liquefied during periods of excess gas availability or imported in liquid form from surplus areas can be stored for extended periods of time in above-ground or underground tanks, as described in Chapter 6. Vaporisation rates, provided insulation is sufficient, can be as low as $0.1\%$ per day of the tank content, although reliquefaction facilities usually cater for about $0.3\%$ boil-off. However, supplementary gas supplies from storage are, as a rule, required in larger quantities and at short notice. Natural evaporation is not capable of supplying these amounts and forced vaporisation in special plant is required.

As an alternative to vaporising the liquid and distributing gas through an existing grid one can, however, make use of the unique properties of high density and concentrated calorific value of LNG and distribute it as such without vaporisation. There are situations where this is advantageous, and road transport, rail cars, barges and even pipelines can be used to carry LNG from storage tank to consumer.

In the present chapter it is proposed to discuss methods and equipment for LNG vaporisation, the distribution of LNG as such, the situations in which this is economically advantageous, and the equipment recommended for the handling of cryogenic LNG. A final section discusses the differences in quality and the degree of interchangeability of various types of LNG and pipeline gases.

## 7.2 REVAPORISATION OF LNG

Since LNG is generally stored for use during peak periods of gas demand, the ability of a gas storage facility to supply large volumes of gas at the shortest possible notice will be its most valuable feature. Design of liquefaction/evaporation plants for peak shaving purposes will, generally, provide limited liquefaction capacity but massive vaporisers to meet sudden peaks of gas demand.[2,7,22]

Vaporisers for LNG are basically heat exchangers[2], the hot fluid being either combustion gases—in direct fired equipment—or steam

93

or hot water—in indirectly heated vaporisers. Very large vaporisers can, in the absence of a source of steam or hot water, also be heated by cooling water. In the latter case a large supply of water, generally river, estuary or sea water, is required so as to avoid excessive cooling of the heat source. In particular, freezing of the cooling water must at all times be avoided, one method being the use of an intermediate heating fluid such as propane.[11]

The usual practice is to pass LNG vertically upwards through pipes embedded in aluminium panels which are suspended from a steel structure. Fresh cooling water is pumped to the top of the panel and runs down from a trough along the side of the panel, to be collected in a pond beneath, whence it is returned to the river or estuary. Internal pressure varies from 7 to 75 atm and plant capacities of up to 60 tons/hr of LNG have been designed. To prevent freezing and ice formation on the tubes 0°C, for sea water of −2°C freezing point, is the lowest permissible exit temperature.[9,27] Open rack panel evaporators, as described, have been built for LNG revaporisation facilities in the UK,[26] France[7] and Japan.[19]

A recent improved design for running film plate exchangers, which is claimed to be more economic in terms of heat flux per unit investment, is based on corrugated stainless steel sheets, spot welded in such a way that once internal gas pressure is applied they will inflate to form pear-shaped channels. The plates are heated externally by cooling water (or an intermediate heat transfer medium) and LNG is vaporised and superheated as it flows horizontally through each heat exchanger pass. Heat fluxes exceeding 50 000 Btu/hr × ft$^2$ × °F (244 000 kcal/hr × m$^2$ × °C) are obtainable with this type of equipment.[27]

An alternative design, the indirect or intermediate fluid vaporiser consists of a steel vessel, which is half filled with liquid propane at a pressure of about 4 atm and a temperature of −1°C. The vessel also contains two heat exchangers, one for sea water, located in the lower half of the vessel, the other for LNG, which is suspended in the vapour space.

In the most common version of such an intermediate fluid vaporiser the LNG heat exchanger is made up of finned aluminium alloy tubing in the shape of double hairpins which are welded into aluminium headers and through which the liquid is passed, while propane vapour condenses on the outside of the tubes and on the fins. The sea water heat exchanger also consists of an aluminium tube bundle, but the tubes are straight and are expanded into mild steel plates which are bolted to manholes at each end of the pressure vessel.[17]

Since the maximum LNG temperature which can be attained in

propane cooled heat exchangers is about $-50°C$ it is generally necessary to heat the LNG vapour further, e.g. by passing it through standard tube and shell heat exchangers with the gas flowing through the shell and sea water in the tubes. Standard practice is to release the gas at $0°C$ and up to 75 atm pressure. Intermediate fluid vaporisers operating in this fashion are in use at the Canvey Island and Le Havre LNG vaporisation plants in the UK and France respectively.

When using a heating medium other than water or propane the temperature difference between heat source and LNG becomes even more pronounced. Heat transfer coefficients are reduced, mainly owing to the formation of a vapour barrier on the LNG side, a phenomenon known as film boiling. The tendency to film boiling can be reduced by increasing turbulence. In the design of so-called high performance evaporators, for instance, it was found that the installation of corrugated multi-fin coils improved heat transfer by a factor of ten.[24]

If the heating medium is hot water a definite improvement in heating efficiency can be achieved by resorting to submerged combustion. By introducing pressurised gas and air below the water surface one can obtain a more rapid heat exchange between combustion gases and water, and furthermore since all the water vapour in the combustion products condenses the gross rather than the net calorific value of the fuel can be utilised. In situations such as vaporisation of LNG where temperature of the heating medium is relatively unimportant, it is clearly possible to use submerged combustion to heat the transfer fluid and exploit its advantages.[9]

On the other hand, LNG revaporisation can be used as a source of cheap refrigeration. A number of cryogenic processes, to be discussed in greater detail in Chapter 8, can utilise the cold produced in the course of LNG vaporisation for such diverse purposes as air separation, shrink fitting of metal parts, solid carbon production, freeze drying, etc. Finally where gas turbines are used as drivers for gas compressors, blowers, etc. their efficiency can be increased by heat exchange between vaporising LNG and combustion air.[9]

## 7.3 DISTRIBUTION OF LIQUEFIED GAS

While LNG is normally gasified before distribution there are certain situations in which it is preferable to ship the gas in liquefied form from a central storage tank to outlying distribution centres.[3] Normally the liquid is transported in insulated containers mounted on truck or rail chassis or barges, but experiments in piping liquid LNG have taken place and proposals for liquid natural gas lines have been made.[6]

### 7.3.1 LNG Pipelines

Work on piping natural gas in liquid form has been in progress for some time now in both Russia and North America.[4,12,13,15,16] The incentive in both countries is the potential reduction in cost of long distance movement of large volumes of natural gas. Both in Siberia and in Alaska and Northern Canada there are large natural gas fields between 1000 and 1500 miles from centres of gas consumption. Transport of gas as such requires not only very large diameter pipe, but also high pressure operation. Furthermore the terrain does not lend itself to buried pipe laying and large temperature changes with corresponding pipe movement complicate above-ground construction.

It is interesting to compare the capacity of pipe of a given diameter and for a given pressure drop for LNG on the one hand and natural gas on the other. For a 36-in pipe with an initial pressure of 55 atm and a pressure drop of 25 atm over 60 miles, throughput of gas in the form of LNG is about $25 \times 10^9$ Nm$^3$/year or 2500 million scfd. The same pipe when carrying natural gas with the same pressure drop over the same distance will have a capacity of only $10 \times 10^9$ Nm$^3$ or 1000 million scfd.[16]

Both liquid and gaseous transport require additional pumping or compression along the line if excessive pressure drop is to be avoided. However, in addition to recompression as in the gaseous phase transport case, LNG must also be cooled repeatedly in order to avoid vaporisation loss and the development of two phase flow, although careful insulation can reduce the need for and increase the spacing of cooling stations. In practice it will be desirable to combine cooling and recompression, i.e. calculate line diameter and insulation such that the same number (or a multiple of) pumping and cooling stations results.

Since LNG lines operate at temperatures of about $-160°C$ the use of mild steel for such lines is no longer possible.[10] Steels of 9% nickel content which do not undergo embrittlement are relatively expensive, and it has been suggested that alloying with 1·5% nickel and 0·1% molybdenum will result in a suitable material which would be no more than 20% dearer than conventional pipeline steels.[16] Canadian authors have proposed aluminium pipe and suggest overcoming the problems of thermal expansion by the use of corrugated pipe. The latter, although it intensifies turbulence and therefore heat generation in the pipe, is preferable to complicated expansion joints and mechanisms for pipe movement which are unavoidable in the case of steel pipe.[25]

The only practical insulation material for LNG lines so far is polyurethane foam. A vapourproof external seal must be applied and protects the insulation against both mechanical damage and water penetration. It will be appreciated that any vapour entering the insulation would ultimately freeze and result in mechanical breakdown, quite apart from the much higher thermal conductivity of ice compared with foam. The thickness of the insulation will determine heat gain of the pipe and its extra cost must be weighed against the additional refrigeration load, in other words the frequency of cooling stations along the line which would otherwise be required.

Materials other than closed cell insulating foams, e.g. polystyrene or expanded perlite, have been considered and, in particular, evacuated jackets, either empty or filled with perlite, could be effective. However, owing to their vulnerability they are less likely to be used than foam. Application of foam insulation can be either *in situ* after completion of the line and before burial or to individual length of pipe at the manufacturers' or in field workshops. In the latter case gaps between insulated sections are filled in with foam prior to line burial.

The principal problems associated with LNG and similar cryogenic lines are due to vaporisation in the line whenever unsteady state conditions prevail. Thus cooling down the line before pumping starts will lead to extensive vaporisation of the initial flow, and gas vents will have to be provided. Similarly turn-down or discontinuing the flow will result in evaporation of the LNG left in the pipe, and again pipe venting will be necessary, the resultant gas losses being unavoidable.

Ideal conditions for a liquefied natural gas pipeline thus appear to be:

— availability of the liquid, i.e. no need to liquefy specially for transport;

— some vaporisation admissible at the delivery end, i.e. no need for recooling;

— short pumping distances, i.e. no need for intermediate pumping or cooling;

— constant flow and consumption of gas, i.e. no turn-down or interruptions.

It will be appreciated that such ideal conditions will be rare, and in consequence wherever the two alternatives of a vapour or liquid line coexist, the former will generally and with few exceptions be preferred on economic, operational and technical grounds.

### 7.3.2 LNG Distribution by Road or Rail Tanker or by Barge

The transport of cryogenic liquids in ship or vehicle mounted tanks is well established and road or rail tankers and occasionally LNG barges are widely used to supply larger gas customers, satellite gas distribution plants and peak load facilities.

The tanks used in such vehicles or ships do not differ greatly from ships' tanks for the ocean transport of LNG. In general a stainless steel (type 304) inner vessel is suspended within an outer carbon steel shell. The annular space between the two vessels is filled with expanded perlite and evacuated; the suspension system, although of stainless steel, is such that only minimum heat leakage occurs. In a current standard US design for a rail car carrying 55 tons of LNG the internal pressure rise over a period of 40 days amounts to only 30 psi. Cryogenic containers mounted on road vehicles (trucks, trailers and semi-trailers) are smaller; in the United States standard LNG trailer sizes go up to 17·0 tons; truck mounted containers of up to 9·0 tons capacity are in use; semi-trailers carry volumes about the same as those carried by vehicles with two or more axles, all LNG transport being size rather than weight-limited owing to the relatively low density of the liquefied gas.[20]

Filling and emptying of LNG tankers can be either by pressurisation or by liquid pumped transfer. In the former case a coil in the tank from which the transfer is to take place is heated by means of steam and the pressure which develops inside the tank permits transfer of the liquid into a receiver. There will be no vapour return connection. On the other hand, if the permissible receiver pressure is low, liquid transfer will require a cryogenic pump; a vapour equalisation line between tanker and receiver facilitates the transfer and maintains pressures throughout close to atmospheric.[14,23]

Railcar, truck or trailer mounted LNG vessels must be provided with fittings for loading and unloading, i.e. fill and drain valves, vapour return valve, pressure relief valve (usually there is a rupture disc in parallel with the latter) and with instruments including a liquid level gauge, pressure gauge and, as a rule, a temperature gauge. A trycock at the level of maximum fill provides a check on the liquid level instrument and protects against overfilling. Relief valves and rupture disc discharge into a vent which should be protected by a flash back arrester in case the escaping gas were to be set alight. While rail cars usually have fittings for loading and unloading at diagonally opposite sides, road vehicles have only one set of valves, as a rule in a cabinet at the rear of the vehicle. In addition, the outer vacuum space also has to be provided with protection against vacuum loss and excess pressurisation due to LNG leakage, usually in the form of pressure gauges and a rupture disc.

If transfer of liquid from the tanker into a receiver is undesirable the possibility of using skid mounted LNG tanks carried by truck to their destination and unloaded by hoist or fork truck should be borne in mind.

Transport in barges was originally proposed for the supply of LNG from Louisiana to the Chicago stockyards. Construction of tankage was similar to that of the first methane ships, i.e. balsa wood on the inside of carbon steel tanks. The concept was not entirely successful and later barge design resorted to insulated stainless steel or aluminium containers of roughly the same dimensions as the ship's compartment in which they were located. Transfer of cargo from the barges to receivers would be either by gas lift from a separate pressure vessel, or by submerged cryogenic pumps.[5]

## 7.4 EQUIPMENT AND MATERIALS

Since carbon steels are unsuitable for cryogenic duty owing to their loss of strength and ductility at low temperatures, tanks for vehicles and ships must be constructed either in low to medium nickel content alloy steels or in various aluminium alloys. Mechanical standards for the various metals used in cryogenic equipment have been codified in Appendix Q of API Standard 620, from where Table 7.1 is extracted.[10]

TABLE 7.1
Allowable Design Stresses for Cryogenic Structures (in psi)

| Spec. number | Minimum strength | | Allowable stress | |
| --- | --- | --- | --- | --- |
| | Tensile | Yield | Design | Test |
| A 353-64 (9% Ni) | 100 000 | 75 000 | 28 500 | 42 000 |
| A 240-304 (stainless) | 75 000 | 30 000 | 20 000 | 24 000 |
| B 209-3003-0 (A1 alloy) | 14 000 | 5 000 | 3 350 | 4 000 |
| B 209-5456-0 (A1 alloy) | 42 000 | 19 000 | 12 600 | 15 200 |

A special grade of treated 5% nickel alloy has also recently been claimed to meet the standard.[8] Clearly, where deposited weld metal has a lower tensile strength than that of the base plate material, or where the latter is affected by the heat of welding, design and test stresses must be reduced accordingly.

Of the materials listed, stainless steels are the most expensive and therefore used only when other metals cannot be substituted; 5 to 9% nickel steel, while somewhat cheaper than stainless steel, is in addition stronger even than the highest tensile aluminium alloy (5456) and also more resistant to corrosion by alkali or acid than the latter. Aluminium alloys are employed where tensile strength is of minor concern and also where the ease of fabrication of aluminium by casting, rolling, cutting etc. is an advantage. Invar, a stainless steel of 36% nickel content, is used only where dimensional change due to temperature changes must be kept to a minimum, its thermal expansion coefficient over the ambient to cryogenic range being almost nil. Mechanical properties of these materials at cryogenic temperatures are further discussed in Chapter 9.

Moving parts in submerged pumps, valve seats, valve spindles and other close fitting metal parts exposed to cryogenic temperature need not generally be made of minimum expansion alloys, provided surfaces in contact are of the same material or have the same expansion coefficient. What must be avoided is the matching of parts from different materials which may show perfect fit at room temperature, but will no longer do so at cryogenic temperatures.

The use of elastomers must clearly be avoided altogether since practically all rubbers lose their elasticity at cryogenic temperatures. Similarly most plastics become exceedingly brittle and should only be used where this is of no consequence. Glass reinforced plastics, on the other hand, can occasionally be used and behave very much like reinforced concrete.

Concrete, while somewhat brittle at low temperatures, can be reinforced by means of prestressing wires, which need not be stainless, to such an extent that it becomes a suitable material for LNG container walls or roofs. No mobile LNG containers have so far been built in concrete, but ship and barge construction has been proposed.

## 7.5 INTERCHANGEABILITY OF LNG AND PIPED NATURAL GAS

The introduction of liquefied imported gas into a distribution grid which has, in the past, only been used to handle piped natural gas, can under certain circumstances result in composition changes of the send-out gas which are of sufficient magnitude to affect its combustion characteristics to a noticeable extent.

While both pipeline gases and LNG are often of the same origin and therefore contain the same impurities, one effect of liquefaction is the concentration of higher boiling components; since the heavier hydrocarbons have a higher calorific value than methane, the chief component of natural gas, LNG prepared from a gas containing ethane and heavier gases will tend to be richer. Similarly if a gas of high nitrogen content, e.g. Groningen gas, is liquefied the LNG will tend to be richer because nitrogen concentration will be reduced in the course of liquefaction. Finally, certain gases produced in association with crude oil will be richer than natural gases, consisting almost exclusively of methane.

Table 7.2 lists gas composition and certain combustion properties

TABLE 7.2

Composition and Combustion Properties of Certain Natural Gases

| | Liquefied natural gases | | | Groningen | Pipeline gases | |
| | Algerian | Libyan (total) | Libyan (fract.) | | Italian | North Sea |
|---|---|---|---|---|---|---|
| COMPOSITION (% VOL) | | | | | | |
| Methane | 86·3 | 66·8 | 80·2 | 81·30 | 95·9 | 85·9 |
| Ethane | 7·8 | 19·4 | 18·2 | 2·85 | 1·4 | 8·1 |
| Propane | 3·2 | 9·1 | 0·3 | 0·50 | 0·4 | 2·7 |
| Butanes | 0·6 | 3·5 | 0·1 | 0·14 | 0·3 | 0·9 |
| Pentanes | <0·1 | 1·2 | <0·1 | 0·07 | 0·1 | 0·3 |
| Nitrogen | 2·0 | nil | 1·1 | 14·35 | 1·8 | 0·5 |
| Carbon dioxide | nil | nil | nil | 0·89 | 0·2 | 1·0 |
| COMBUSTION PROPERTIES | | | | | | |
| Gross cal. value (kcal/m$^3$) | 10 750 | 13 367 | 10 712 | 8 460 | 9 900 | 10 570 |
| Modif. Wobbe Index | 13 375 | 14 914 | 13 287 | 10 550 | 13 000 | 13 050 |
| Combustion potential | 44·0 | 46·2 | 49·2 | 35·6 | 43·0 | 43·8 |

for a number of liquefied gases of different origins and also certain types of gases distributed in Europe. The combustion characteristics quoted are the gross calorific value, the modified Wobbe Index, a measure of the thermal output of an atmospheric burner operating at constant gas pressure, and the Delbourg Combustion Potential, a function of the velocity of flame propagation in the gas. For two gases to be fully interchangeable their Wobbe Indices should not differ by more than 5%—a fluctuation of 10% would be acceptable in emergencies—and the typical combustion potential of about 35 for most natural gases should never exceed a value of 50. Calorific value of a substitute gas is not of particular significance as far as technical interchangeability is concerned; however, commercial considerations force a gas distributor to maintain at the very least a constant average heating value.

It will be gathered from Table 7.2 that the combustion properties of 'rich' natural gases such as North Sea Ekofisk or Italian Cortemaggiore gas can be matched more easily by LNG than can those of Groningen gas. In fact before introducing LNG into a distribution system dedicated to the latter it will always be necessary to condition the supplementary gas by blending with an inert component. In the case of Cortemaggiore gas, on the other hand, some associated rich gases, e.g. Libyan LNG, will require conditioning, while others such as Algerian gas could be blended up to a reasonable proportion without prior treatment.

It follows that whenever LNG is revaporised and sent out in gaseous form in a distribution system which is not exclusively used for regasified LNG, the combustion characteristics of both gases must be determined and if necessary, their properties adjusted to ensure complete interchangeability.

## REFERENCES

1. Ambler, B. M. (1967). *Over-the-Road Transportation of Liquid Methane*, A.G.A. Distribution Conf., St. Louis, May 1–4.
2. American Gas Association (1968). *LNG Information Book*, Section IV, New York.
3. American Gas Association (1968). *LNG Information Book*, Section V, New York.
4. Anderson, J. H. (1965). Liquefied methane pipeline, *Oil Gas J.*, 63(6), 74–80.
5. Biederman, N. P. (1972). LNG barges may solve many problems, *Pipeline Gas J.*, 199(7), 47–52.
6. Bloembaum, Jr. W. D. (1968). *Transportation of LNG for Remote or Interruptible Service*, Paper 40, LNG–1, Chicago.

7. Buffiere, J. P. (1969). *LNG Terminals—Description of Vaporisation Processes and Corresponding Materials*, Proc. Int. Conf. LNG, pp. 625–642, London.
8. Cordea, J. N., Frisby, D. L. and Kampschaefer, G. E. (1972). Steels for LNG storage, transportation, *Oil Gas J.*, **70**(41), 85–90.
9. Crawford, D. B. and Eschenbrenner, G. P. (1972). Heat transfer equipment for LNG, *Chem. Eng. Prog.*, **68**(9), 62–70.
10. Dainera, J. (1968). *Considerations for LNG Pipe Material Selection*, Paper 42, LNG–1, Chicago.
11. Dale, S. E. (1971). A new look at LNG vaporisation methods, *Pipe Line Ind.*, **35**(1), 25–28.
12. Dimentberg, M. (1967). *Transmission of Natural Gas in Liquid Phase by Long Distance Pipeline*, Cryogenic Engineering Symposium, Toronto, June 4–7.
13. Duffy, A. R. and Dainera, J. (1967). LNG pipelines appear technically feasible, *Oil Gas J.*, **65**(19), 80–89.
14. Eiffel, P. J. (1968). *Railroad–Highway Transportation of LNG*, Paper 38, LNG–1, Chicago.
15. Gibson, C. J. (1968). *Feasibility of Large Super Insulated Pipelines for Transport of LNG*, Paper 41, LNG–1, Chicago.
16. Gudhof, S. F. (1969). *Pipeline Transmission of Natural Gas in Liquefied and Cooled Gaseous State*, Proc. Int. Conf. LNG, pp. 544–559, London.
17. Guthrie, J. K. and Gregory, E. J. (1969). *Design of Base Load Evaporators for LNG*, Proc. Int. Conf. LNG, pp. 300–321, London.
18. Herve, P. (1968). *The Production of LNG for Peakshaving and the Transport of LNG by Pipeline*, Paper 29b, LNG–1, Chicago.
19. Hagiwara, Y. (1968). *Specific Features of the Yokohama LNG Receiving and Utilisation Facilities*, Paper 14, LNG–1, Chicago.
20. Latham, W. N. (1969). *The Design of Optimum LNG Highway Tankers*, Proc. Int. Conf. LNG, pp. 592–606, London.
21. Maishman, W. G. and Potter, J. H. (1969). *Transport, Temporary Storage and Vaporisation of LNG*, Proc. Int. Conf. LNG, pp. 560–580, London.
22. O'Donnell, P. J. (1971). Gas Council using LNG for peak shaving, *Oil Gas J.*, **69**(40), 94–96.
23. Simpson, G. E. (1964). Cryogenic transport design considerations, *Cryogen. Technol.*, **1**(5) 27–30.
24. Waldmann, H. (1969). *Development of a High Performance Evaporator for Liquefied Gases*, Proc. Int. Conf. LNG, pp. 322–341, London.
25. Walker, G., Coulter, D. and Sood, N. (1969). *LNG Pipelines for Arctic Gas Recovery*, Proc. Int. Conf. LNG, pp. 503–523, London.
26. Ward, J. A. (1968). *Latest Developments at Canvey Island*, Paper 17, LNG–1, Chicago.
27. West, H. H., Hashemi, H. T. and Wesson, H. R. (1972). *Design of Running Film Plate LNG Vaporizers*, ASME Paper 72-Pet-34, New Orleans.

*Chapter 8*

# Utilisation of LNG

## 8.1 INTRODUCTION

Normally LNG from bulk storage facilities is distributed either as such, or after regasification, and used by local gas undertakings in their public distribution system. In particular, all peak shaving LNG is handled in this fashion, and at present the bulk of LNG supplies is still earmarked for this particular end use.

Only relatively small volumes of LNG, at the moment anyway, find their way into outlets other than public gas distribution. Direct LNG applications, which have recently been studied and some of which have been commercialised, include its use as a transport fuel, i.e. replacement for motor spirit in spark ignition engines, its use in other prime movers such as gas engines and gas turbines, the latter both on the ground and in the air. In particular, LNG as a supersonic jet fuel and as a fuel for helicopters needs mentioning.

Another aspect of LNG utilisation is worth reporting. The extreme cold resulting from LNG vaporisation can be used for various cryogenic technologies, either in conjunction with and as a by-product of LNG vaporisation for public distribution, or in a self-contained process which requires both refrigeration and a gasified fuel. An instance of the former is the combination of air separation and oxygen manufacture with LNG regasification. An example of a self-contained process is the refrigeration of the cargo space of trucks and lorries with LNG, prior to use of the vaporised gas as an engine fuel.

A technique of potential significance is the exchange of cryogenic temperatures between LNG and a refrigerated carrier fluid or cryophore, which in turn can be used to refrigerate tankage and precool additional quantities of natural gas. Nitrogen can, for instance, be liquefied with the aid of LNG and can be shipped back to the liquefaction plant to maintain vessel tanks at low temperature and to precool fresh gas before liquefaction. Other cryophores are being considered.[15]

## 8.2 LIQUEFIED NATURAL GAS AS A FUEL

The main characteristics of LNG which have a bearing on its performance as a fuel are its low temperature and clean burning.[11] Liquid methane also has high knock resistance in spark ignition engines and burns smoother in compression ignition engines. It is therefore of interest as an automotive and aviation fuel.

### 8.2.1 Automotive Uses

Since LNG burns to carbon dioxide and water, even under conditions when other fuels would form undesirable partially oxidised and sometimes odorous compounds, it has been claimed widely that noxious emissions from motor car engines could be reduced, if not eliminated, by switching from gasoline to natural gas as fuel. Apart from lower emission of soot, aldehydes, unsaturates, carbon monoxide and similar intermediates formed during hydrocarbon combustion, natural gas which is free of sulphur does not form sulphur dioxide on oxidation and also, unlike leaded motor fuels, does not emit air polluting lead compounds.[4,8]

While the emission of hydrocarbons from LNG fuelled engines is not zero, it can be shown that most of the emitted material is paraffinic; it is mostly unburnt methane, and contains little if any photochemically reactive olefins or acetylenes. Similarly the other objectionable component of exhaust gases, nitrogen oxide, while not eliminated by a change to LNG, can be substantially reduced by a change in engine operating conditions. Lean mixtures form less $NO_x$, and methane fuelled engines can be operated much leaner with little or no surging, rough running or poor idling compared with gasoline engines.

Reduction or elimination of carbon monoxide can also be achieved but depends on careful proportioning of air and gas. A variable venturi mixer allegedly has been found effective and virtual elimination of carbon monoxide has been claimed.[8]

Inevitably there are certain disadvantages associated with the use of LNG as an automotive fuel. Maximum engine power is reduced if a gas rather than a liquid is introduced with the combustion air. Peak cylinder pressures are also reduced, owing to the slower flame speed of methane, and power output is lowered further unless spark timing is advanced. In a system adjusted for optimum emission of pollutants this power loss amounted to about 15%.

LNG tanks holding fuel of the same heat capacity also occupy a greater volume than their gasoline counterparts, not only owing to the lower density of LNG but also because they have to be surrounded by an evacuated insulation space. The cost of conversion to dual fuel operation, and in view of the relative scarcity of LNG refuelling points dual fuel facilities are essential, will add a small but not insignificant amount to the first cost of the vehicle, and while the risk of carrying LNG is probably no worse than that of carrying gasoline, insurers and licensing authorities will take time to fully accept the new fuel.[2,12]

The future for LNG fuelled motor cars, therefore, depends on the cost differential between the two fuels; if differential taxation, either as a direct fuel tax, or in the form of a pollution tax, encourages conversion, LNG fuelled motor cars may become more than a technological curiosity.

The use of LNG to fuel diesel engined trucks provides an opportunity of not only reducing air pollution by exhaust gas but also to refrigerate perishable cargo by first evaporating LNG in a heat exchanger. Experimental refrigerated trucks have been designed, and dual fuel operation is again the preferred mode. While liquid fuel in a diesel engine is injected, natural gas must be aspirated together with combustion air. Methane has the necessary knock resistance to replace most of the injected fuel, but owing to its unsatisfactory ignition characteristics (cetane number) cannot be used for start-up.

Diesel buses have been converted to dual fuelling mainly on grounds of air pollution. Since they tend to operate in thickly populated areas clean exhaust conditions are often of paramount importance. Fuel tankage and its protection against impact, cold starting and acceleration problems do not appear to be insoluble.

A frequently voiced objection to LNG fuelling, the LNG loss resulting from interrupted use of the vehicle, is probably less significant than it sounds; LNG vacuum jacketed tanks incur minimal LNG losses during 72 hr and even if a motor car is garaged for this period fuel loss will be small and normal ventilation of a garage capable of dealing with the escaping gas.

### 8.2.2 Use in Aircraft

While LNG used as a fuel in surface transport provides the obvious advantages of clean emission and improved engine operability, the most important LNG characteristic to be utilised in aviation is its low temperature and latent heat of vaporisation.[7]

Skin friction, particularly in supersonic flight, produces substantial quantities of heat and in order to protect the structure of supersonic transport aircraft it is necessary to use expensive materials of construction for all parts of wing and fuselage which are subject to frictional heating. One way of avoiding the use of stainless steel sandwich or titanium construction is to refrigerate the areas in question, and it has been proposed to circulate LNG in such a way that overheating of leading edges and other exposed areas is prevented. The LNG vaporised by frictional heat is subsequently used as fuel in the propulsion unit.[13]

In addition, LNG fuelling of a gas turbine confers certain benefits. Owing to its higher heating value per unit weight (about 13% more than ordinary jet fuels) weight-limited aircraft can be given additional operating range or increased carrying capacity. This does not, however, apply if limitations are volumetric; the density of LNG being some 45% less than that of jet fuel, tankage for LNG to give the same range as jet fuel will have to be correspondingly larger.

A more interesting way of using LNG as a heat sink, which applies to both subsonic and supersonic aircraft, is the possibility of cooling hot engine parts in gas turbines. If turbine inlet temperatures—at present of the order of 820°C—could be raised to 1650°C, by cooling turbine blade tips and combustor cans with LNG for instance, this would more than triple power output of the engine at compression ratios of 8–10. If simultaneously compression ratio were to be increased to 25 the engine horsepower per pound of air consumed could be quadrupled and its efficiency increased by at least 50%.

Alternatively, rather than raise engine performance and efficiency, one can improve engine life by cooling turbine blades and combustors without any increase in operating temperature. A further contribution to engine life and reduced maintenance is made by the cleaner burning characteristics of methane compared with jet fuels. The absence of deposited carbon, which tends to distort combustors, emits thermal radiation, and can build up into solid particles which may become detached and damage turbine blades, is the result of greater chemical stability and reduced thermal cracking of methane compared with jet fuel. Finally the absence of sulphur and other non-hydrocarbon elements in LNG also mitigates corrosion of the gas turbine and its downstream equipment.

Particular advantages are claimed for LNG as a fuel for helicopters and turboprop engines.[20] In both instances LNG, in addition to refrigerating turbine blades, combustors and other engine parts, can be used to cool the power transmission train, the lubricant of which invariably has to be cooled to ensure satisfactory operation.

Prospects for the use of LNG as an aircraft fuel depend to some extent on the availability of suitable lightweight fuel tanks. While LNG would be circulated through the leading edges of SST wings etc. during supersonic operation, both on the ground and during subsonic flight it will have to be stored in lightweight, gasproof and thermally insulated fuel tanks. Double skinned vacuum jacketed tanks will generally be too heavy for aircraft use—they would weigh about the same as the fuel itself in the 400 to 4000 litre range and about two-thirds in larger sizes. Foamed plastics with gas-impermeable internal membranes and external sealing against water vapour would be much lighter—about 15% of the weight of the fuel. However, they will have to withstand repeated thermal contraction and expansion.

An interesting argument in favour of LNG compared with jet fuels is that of safety. Owing to its low boiling point LNG does not form inflammable mixtures in the ullage space of tanks even if they are ruptured. Ignition temperatures of methane/air mixtures are also comparatively high, about 650°C, compared with that of kerosine/air, which is 260°C. There is also little danger of electrostatic ignition during LNG refuelling whereas in kerosine entrained water and rust particles can easily raise electrical conductivity to the danger level of 1 to 10 picomho/metre. Finally when spilt LNG not only cools hot metal, thus reducing the danger of ignition, but also evaporates and disperses completely within a short period.[1]

There are, of course, certain drawbacks associated with LNG as an aviation fuel. Chief among these is the cost of converting existing refuelling facilities, fuel tanks and fuel supply to LNG in a sufficient number of aircraft and at enough airports to make the use of LNG a practical proposition. Even for supersonic airliners of the future the use of aviation kerosine rather than LNG has prevailed and neither the Concorde nor the Russian SST will fly on LNG. Other disadvantages are the need for a return vapour line while refuelling, the extra weight of LNG tanks and the public concern about any novel and as yet untried fuel, particularly a cryogenic material such as LNG.

## 8.3 COLD UTILISATION

The extensive use of refrigeration techniques in industry and commerce presents the user or distributor of LNG with an excellent opportunity to utilise the by-product refrigeration which results from the regasification of liquid natural gas.[17] Refrigeration, as discussed under the heading of Gas Liquefaction in Chapter 4, is expensive and demanding in mechanical energy and capital investment. Savings in both can be made by substituting the cold generated during regasification for the electric or other energy required to drive compressors and also by eliminating all refrigeration equipment other than heat exchangers.

The main problem of LNG utilisation is the seasonal nature of LNG revaporisation. Since peak shaving LNG plants do not require gas for most of the year their revaporisation cold capacity is only available for a limited period. Most industrial and other users of refrigeration would therefore have to maintain stand-by plant if they were to rely on peak shaving LNG.[6,18,19]

Revaporisation of imported base load LNG does not suffer from this drawback, although even here some schemes may be designed for higher winter and lower summer revaporisation rates. Their main problem is the large quantity of refrigeration available from any one facility. Only very large users of cold such as air liquefaction and separation plants, ethylene or hydrogen plants, very large cold stores, some frozen food manufacturers and possibly freeze purification plants for saline water or sewage will find a use for the massive quantities of cold generated in such a plant.

### 8.3.1 Liquid Air Plants

The liquefaction of air as a rule serves the ultimate purpose of air separation into nitrogen and oxygen, sometimes accompanied by the separate recovery of argon. If all the product streams are recovered in gaseous form heat can be exchanged between warm incoming air and the vaporised cold individual gases, and refrigeration requirements for the separation will be accordingly small.

In actual fact some of the products, particularly oxygen, are often needed as liquid in order to facilitate transport and storage, and under these circumstances further non-recoverable refrigeration is required.[9] Nitrogen, similarly, may be used in liquid form for refrigeration or deep freezing of food and other perishable goods or for the manufacture of ammonia, where liquid nitrogen is used to condense residual carbon dioxide and other impurities from the hydrogen stream before pure hydrogen and nitrogen are reacted. In either case it can no longer be used for heat exchange with the incoming air and further external refrigeration must be supplied.

A special method of utilising liquid nitrogen produced in an air separation plant, which forms part of a liquid natural gas terminal, is to ship the 'cold' back to the LNG liquefaction plant overseas.[21] Air is liquefied and separated into liquid nitrogen and either gaseous or liquid oxygen. There are a host of outlets for the latter in steel making, chemicals and engineering, but disposal of large quantities of liquid nitrogen can be difficult. The liquid is therefore loaded into the cryogenic tankers after they have discharged the LNG and is shipped back to the place of origin of the gas, thereby cooling the ship's tanks during the return voyage and also providing an additional source of refrigeration for further natural gas liquefaction at its destination.

While the utilisation of the refrigeration due to LNG vaporisation by return shipment of liquid nitrogen depends on the availability at close quarters of an air separation plant, there are no such restrictions if cryophores other than nitrogen are used. A method which was recently described recommends the use of a non-disposable cold carrier, e.g. isopentane, which is transported back to the liquefaction plant at cryogenic temperatures and, after heat exchange with fresh natural gas feed, is transported at ambient temperature forward again to the LNG vaporisation facility.[5]

Techniques for the liquefaction of air are well developed and well known. It is therefore not proposed to discuss them in detail beyond stating that, much as in natural gas liquefaction, Joule–Thompson or Claude (expander) cycles are used, with Stirling cycles being under consideration for larger, but so far confined in practice to the smaller, units. Heat exchange, both in regenerative and recuperative equipment, is employed to precool the incoming gas, and indirect heat exchangers are, obviously, essential if LNG vaporisation is to supply part of the required refrigeration without mutual contamination.

Air purification before liquefaction involves the removal of water, $CO_2$, $H_2S$ and hydrocarbons, the latter because of explosion danger in the presence of liquid oxygen. The availability of LNG provides the opportunity to purify by stepwise refrigeration; LNG cooled, paired heat exchangers collect the impurities as solid frost, hydrates and dry ice; after a period of running the exchanger is taken out of operation and the contents are thawed and removed by a stream of warm air. Various other arrangements, depending on product utilisation, are possible to utilise the cold and simultaneously remove the frozen solids content of the exchangers.

**8.3.2 Other Gas Separations**

In addition to separating the components of atmospheric air, cryogenic processes are also used to produce pure streams of hydrogen, ethylene, the other lower olefins and helium.

While air separation is normally based on complete liquefaction and fractional distillation of the liquid this is not normally the case when hydrogen is purified. The low atmospheric boiling point of the latter, and even more so that of helium, permit one to condense all the other gases present and to withdraw pure hydrogen or helium in gaseous form.

The separation of ethylene from thermally or catalytically cracked ethane can take the form of a complex LNG regasification plant in which first ethane is separated from methane and heavier hydrocarbons by fractionation, pure methane being distributed as gas and the heavier materials used as LPG.[16] Purified ethane is cracked and produces a mixture of hydrogen, ethylene, ethane and heavier products formed by polymerisation and recombination of radicals. If this material is cooled and compressed all the gaseous components with the exception of hydrogen can easily be condensed by using LNG as a refrigerant. Fractionating the condensate subsequently results in pure component streams, particularly pure polymerisation grade ethylene.[10]

Other olefins, e.g. propylene, n-butene, and isobutylene are obtained by fractional separation of so-called 'cracker light ends', the vapour fraction produced by thermally or catalytically cracking a hydrocarbon boiling in the range between LPG and gas oil in the presence of diluent steam. Although boiling points of the gases are close, fractional separation either by condensation or by distillation of a mixed condensate is possible. Vaporising LNG can be used to supply the necessary refrigeration for such separations, and the scale of these operations in a modern petrochemical plant is sometimes sufficient to absorb the bulk of the cold available from a base load LNG scheme.

Refrigeration in petrochemical feedstock plants and petroleum refineries is not only required to effect low temperature gas separations, but also assists in the storage of propane and butanes under atmospheric pressure. Insulated tanks are preferred to pressure vessels for storage capacities greater than about 1000 $m^3$, and refrigerated LP gas storage can thus provide additional outlets for the cold available from methane regasification.

The separation of coal gas into hydrogen, generally for ammonia manufacture, ethylene for chemicals, and residual fuel gas, while once an important technology involving intensive refrigeration, has to all intents and purposes been superseded by petroleum based hydrogen and ethylene production. The manufacture of hydrogen from coal or coke by the steam–oxygen reaction, on the other hand, has had a new lease of life as part of the new synthetic or substitute natural gas manufacturing processes, which are based on coal. Again the relatively pure hydrogen required for coal gasification can be separated cryogenically from other gases, and LNG is a suitable source of refrigeration.

### 8.3.3 Miscellaneous Cryogenic Processes

As implied previously, the difficulty of using LNG refrigeration for the host of cryogenic applications which are encountered in modern technology is that total refrigeration requirements rarely fall in line with the cold available from even quite a small LNG terminal.

Many chemical polymerisation processes, for instance, are accelerated by low temperatures, but only rarely does one have to lower the temperature to − 100°C or less, and few polymer plastics are made in large enough plants. One instance of large scale production is the manufacture of butyl rubber, which results from the co-polymerisation of isobutylene and isoprene both dissolved in methyl chloride at about $-40°C$. Other polymerisations and co-polymerisations which require refrigeration are the manufacture of high-density polyethylene, polypropylene and the polyisoprene and polybutadiene rubbers.

Apart from its use in the chemical industry, massive refrigeration is increasingly used in the preservation and storage of foodstuffs. The rapid growth in frozen food sales all over the world has resulted in the large scale manufacture of deep frozen meat, vegetables, poultry, soft fruit and numerous part-prepared dishes and meals. All these are produced by the rapid cooling of the raw or part-cooked food, often by spraying it with liquid nitrogen. Since the cells of animal and plant tissues would burst, and this would affect both texture and taste of the frozen foods, if large ice crystals were allowed to form, it is essential to cool food so rapidly over the range of ice formation ($-1°C$ to $-40°C$) that little or no ice is formed; frozen food packages can subsequently be reheated to their normal storage temperature (about $-10°C$) without further ice crystal formation. LNG is not as a rule used for the direct cooling of foodstuffs, but its cold can be exchanged with nitrogen, the latter being sufficiently inert to be left in long-term contact with food.

Apart from food freezing and the manufacture of cold foods such as ice creams or sherbets, low temperatures are also required for the storage of meat, dairy products and fruit. Refrigerated warehouses will be found in practically all seaports and most larger cities. Temperatures are maintained only a few degrees below freezing point and most fresh foods can be stored under these conditions for several months. While the refrigeration requirements of large food storage complexes are not necessarily out of line with the amount of cold available from a typical terminal the temperature level is clearly too high for optimum economics.

An interesting use of cryogenics occurs in the generation of electricity. The phenomenon of superconductivity, i.e. rapidly decreasing electrical resistance of a metal when approaching a temperature between 1 and 10 K, is well known and can be used to produce unusually strong magnetic fields which are of great value in electricity generation and voltage transformation. While the temperature of LNG boiling at atmospheric pressure is too high to achieve superconductivity in electrical conductors it is possible to use LNG to precool nitrogen or hydrogen, which can be liquefied and evaporated at reduced pressure, to produce superconductivity in suitable metals. Volumes of LNG to be used for superconductivity work would, of necessity, be small, although the low temperature required would provide scope for the efficient utilisation of LNG refrigeration.

A somewhat similar suggestion has recently been made to ship LNG by pipeline and electricity through the same conduit.[14] The location of some power generation plants and LNG facilities is such that some consideration could be given to the construction of common transmission lines; the low temperature of the conductor would ensure that electrical resistance was substantially reduced—the resistivity of copper at ambient temperature, for example, is 2 micro-ohm cm whereas at −160°C it is only 0·45—and heat losses during power transmission would be substantially reduced. Admittedly most of the residual heat generated by the current would be absorbed by the LNG, part of which would consequently evaporate.

Similar relatively small quantities of refrigeration are also used in the engineering industries. It has recently been found that heat treatment of certain steels is improved if a work piece is rapidly chilled (precipitation hardened) to −90°C. This results in an almost complete transition of austenite into martensite (97% compared with 85% at 35°C) and in improved hardness, toughness and ductility. The treatment is recommended for high speed drills and hacksaw blades.

Certain stainless steels (e.g. type 347) of increased tensile strength are obtained by rolling at low temperature (zerolling) and subsequent heating to 430°C. Similarly, explosion forming at low temperature (cryogenic blasting) is claimed to give improved hardness and better penetration of hardness into the sheet. Welds can also be strengthened by this technique. Finally cryogenic cooling of work pieces during machining, particularly in the case of the very hard and high melting 'superalloys' is desirable in order to extend tool life and prevent work hardening and oxidation. Among non-ferrous metals aluminium can be stress-relieved (normalised) by exposure to −90°C for 2 to 3 hr. In all these instances LNG can be used in lieu of nitrogen if processing takes place in a well sealed chamber; alternatively LNG can be used to precool nitrogen or an inert gas mixture which is then used in the low temperature treatment of the metal.

Other potential uses of cryogenics which should be considered are such diverse applications as the use of refrigeration to effect repairs in filled pipelines (two blockages are created upstream and downstream of the repair by freezing the content); the freeze drying and pulverising of foods and chemicals; the shrink fitting of metal inserts into tight fitting outer parts; the use of cryogenic fluids to stabilise soil during excavation; the use of refrigeration to increase the power output of thermal generators and external combustion engines; and many others.

# REFERENCES

1. Anderson, J. K. (1969). *Ground Refuelling System for Liquid Methane Fuelled Supersonic Aircraft*, Proc. Int. Conf. LNG, pp. 678–689, London.
2. Anon. (1967). Clean burning LNG appears economic as a fuel for cars, *Oil Gas J.*, **65**, 214/5, Dec. 25.
3. Buffiere, J. P. and Vincent, R. (1972). *LNG Cold Recovery and Gas Adjustment at the Fos-sur-Mer Terminal*, Session III, Paper 8, LNG–3, Washington.
4. Engler, M. R., Jr. (1968). *LNG Utilisation—SDGE Opens New Fields*, Paper 43, LNG–1, Chicago.
5. Hendal, W. P. and de Grass, J. D. (1972). *LNG Transportation Schemes Based on the Use of Non-disposable Cold Carriers*, Session IV, Paper 3, LNG–3, Washington.
6. Johnson, P. C. (1968). *Satellite Peak Shaving Systems*, Paper 18, LNG–1, Chicago.
7. Joslin, C. L. (1968). *LNG Usage in Aircraft Engines*, Paper 44, LNG–1, Chicago.
8. Karim, G. A. (1969). *Some Aspects of the Utilisation of LNG for the Production of Power in Internal Combustion Engines*, Proc. Int. Conf. LNG, pp. 660–673, London.
9. Katarka, H., Fujisawa, S. and Inone, A. (1970). *Utilisation of LNG Cold for Production of Liquid Oxygen and Liquid Nitrogen*, Session VI, Paper 3, LNG–2, Paris.
10. Kniel, L. (1969). *Utilisation of the LNG Cold Potential (Ethylene Manufacture)*, Proc. Int. Conf. LNG, pp. 643–659, London; also US Patent 3,548,024, Dec. 15, 1970.
11. Maglan, J. E. (1968). *LNG—A Future Fuel for the Transportation Industry*, Paper 46, LNG–1, Chicago.
12. Martindale, D. L. (1971) System considerations for vehicles using LNG, *Cryogen. Technol.* 7(3), 73–75, 87.
13. Matton, G. E. J. (1970). *Analysis of Nominal Characteristics of Liquid Methane SST Turboreactors*, Session VI, Paper 6, LNG–2, Paris.
14. Pastuhov, A. and Ruccia, F. (1970). Why not transport electricity and LNG at the same time? *Pipe Line Ind.*, **32**(5), 44–48.
15. Petsinger, R. E. (1969). *New Applications for LNG*, Proc. Int. Conf. LNG, pp. 75–88, London.
16. Rigola, M. (1970). *Recovery of Cold in a Heavy LNG*, Session VI, Paper 1, LNG–2, Paris.
17. Seliber, J. (1968). *Cold Utilisation in LNG Peak Shaving Plants*, Paper 45, LNG–1, Chicago.
18. Snyder, J. A. (1969). *A Second Generation LNG Peak Shaving Cycle*, pp. 221–230, Proc. Int. Conf. LNG, London.
19. Stark, V. and Wastie, A. E. (1968). *LNG for all Utilities and Satellite Plants*, Paper 47, LNG–1, Chicago.
20. Stephenson, J. M. (1968). *LNG as a Fuel for Helicopters*, Paper 48, LNG–1, Chicago.
21. Williams, V. C., Simmonds, O. H. and Williams, W. M. (1969). *A New Economic Process for LNG Shipping, Liquefaction and Regasification Cold Use*, Proc. Int. Conf. LNG, pp. 487–502, London.

*Chapter 9*

# Safety of LNG Installations

## 9.1 INTRODUCTION

The complete novelty of handling a supercold liquid fuel originally resulted in some hesitation on the part of gas distributors and consumers to become involved with LNG when it was first mooted as a peak shaving aid in the late 1930s. Nevertheless the obvious economic advantages of liquefying surplus gas during the summer and introducing it into the system in the winter soon outweighed these apprehensions, and the first LNG scheme came into operation in the US in 1940 at Cornwell and was soon followed by a much larger facility in Cleveland. This installation was enlarged to hold a total of 270 million scf of gas in the form of LNG. However one 100 MMscf tank failed in 1944 leading to a disastrous fire.[12] Partly as a result of this experience no LNG facilities were built in the US for about 12 years, and even now safety considerations dictate type and location of LNG storage facilities to a much greater extent than for other fuels. Safe storage of LNG and public awareness of its practicability are therefore of considerable importance.[10]

The safe handling of any fuel clearly depends on an accurate knowledge of its characteristics; in the case of cryogenic and similar materials of extreme properties there must be added detailed information on the effect the fuel exerts, owing to its low temperature or other features, on the physical properties of all materials of construction for tanks, pipes, valves, instruments, insulation, etc. Cryogenic substances in particular will produce changes in the physical strength of metals and other materials which can be of serious consequence in regard to safety.

## 9.2 CHARACTERISTICS OF METHANE AT CRYOGENIC TEMPERATURES

Although methane in liquefied form has been known for 50 years or more there still exists a measure of uncertainty and argument as to some of its physical properties. This is mainly due to its cryogenic nature which makes all forms of inspection and measurement somewhat doubtful. Because it is so close to its boiling point, carrying out determinations of physical properties may itself result in a phase change with consequent changes in characteristics.

It seems agreed, however, that liquid methane is a clear, transparent, almost odourless fluid of low viscosity. Its main physical properties[15,19] are listed below:

| | |
|---|---|
| Density at $-161 \cdot 6°C$, 1 atm, g/cm$^3$ | 0·432 |
| Molecular weight | 16·04 |
| Boiling point, 1 atm, °C | $-161.6$ |
| Critical pressure, atm | 45·8 |
| Critical temperature, °C | $-82.1$ |
| Critical density | 0·415 |
| Latent heat of vaporisation at $-161 \cdot 6°C$, kcal/kg | 121·7 |
| Specific heat of liquid at $-161 \cdot 6°C$, kcal/kg | 0·925 |

Methane gas in the region of its boiling point has a specific gravity with respect to air of 0·554 and specific heats at constant pressure and constant volume of 0·5271 and 0·403 kcal/kg × °C respectively. A full list of properties of methane and other gases present in LNG will be found in Appendix H.

Mixtures of methane and air are inflammable over a range from about 5% to about 15% by volume of the gas in air at room temperature, and, as one would expect, the inflammability limits are similar if cryogenic gas is mixed with ambient air.

Other hydrocarbon liquids appear to be freely miscible with liquid methane, and apart from some separation due to fractionation, mixtures of all the lower hydrocarbons of practically any composition can be handled as LNG. However, polar compounds are not normally soluble in liquid methane and this presents problems in regard to gas odorisation, which should always follow rather than precede gas liquefaction.

Water vapour in contact with liquid methane instantly condenses and ultimately freezes. LNG spills will consequently be surrounded by a cloud of condensing steam and, particularly, spills at sea will produce a number of unusual phenomena which will be discussed at greater length. Cooling of atmospheric air and condensation of water vapour will also contribute to the stability and persistence of methane clouds due to spills on both land and sea.

The density of liquefied natural gas is a function of both temperature and composition. Figure 9.1, for example, is a plot of liquid density against heating value and/or average molecular weight of the liquefied gas. It is, of necessity, empirical since gas composition varies not only as far as the total content of higher hydrocarbons is concerned but also in regard to their distribution. However, once one eliminates the effect of nitrogen and other inert gases, a clear correlation exists for the great majority of LNG mixtures.[27]

The significance of liquid density variations in different gases and of changes taking place in the course of vaporisation will be discussed under the heading of stratification and roll-over, phenomena which can occur when different types of LNG are stored in the same tank. Since sudden mixing of two or more blends of different composition

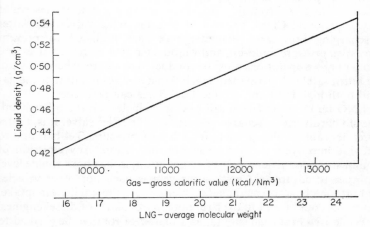

Ｘ Fig. 9.1 LNG specific gravity as a function of composition, assuming a nitrogen content of 2% or less.

or temperature can result in the rapid vaporisation of large quantities of liquid with consequent pressure build-up, differences in specific gravity can be of great significance in regard to the safe operation of LNG facilities.

The low density of methane gas compared with air, even at cryogenic temperatures, facilitates dispersal of gas which has accidentally been released; on the other hand it indicates a special approach to the purging of storage vessels before and after use. Thus nitrogen to replace methane vapour would be introduced near the bottom, whereas methane gas should be charged near the roof. If the vessel is to be cooled down simultaneously LNG would be injected at a controlled rate to prevent sudden chilling of the walls.

The light methane gas will also tend to accumulate in unvented sewers and sewage chambers and under the roof of enclosed buildings.

Well designed ventilation wherever spillage is likely to occur and careful sealing of sewers are therefore absolutely essential. If liquid is spilt from above-ground tanks owing to accident or maloperation the spill must be contained by dikes or drainage and led into a safe impounding area. Volume of the latter must be sufficient to contain the entire content of the damaged tank, and the distance of the impounding basin from potential sources of ignition, neighbouring properties and buildings must be such that there is no danger of a fire. Minimum spacing has been set out by the NFPA[23] and IGE,[17] as discussed in Chapter 6.

It is not possible to confine liquid methane in a pressure vessel and fixed tankage is invariably designed for only a few inches water gauge pressure. Check valves and pressure relief valves, the latter discharging into large diameter pipe, are designed to prevent excessive pressure build-up, and vapours emitted by relief valves and from purging operations must be conducted to a flare, where they can be burnt safely. A permanent small flame ensures ignition of the gas and both high level and ground level flares can be used.

LNG tanks mounted on rail or road vehicles, on the other hand, should hardly ever discharge gas through the relief valve. It is, therefore, desirable to design them for a higher pressure, 3–4 bar, and to provide improved, generally vacuum, insulation. Any emission of combustible gas, if genuinely unavoidable, must be above engine level in order to ensure that neither the vehicle itself nor other vehicles aspirate methane or methane/air blends into the engine air intake, since this could result in over-running, particularly of diesel engines.

While LNG tankage should, as a matter of routine, be earthed to avoid the risks of lightning or other electric discharge, and during transfer of LNG from one tank to another the vessels should be electrically connected, there is in fact less danger of a build-up of high voltage due to liquid flow for LNG than in the case of higher boiling petroleum fuels containing non-hydrocarbon impurities. LNG is an excellent insulator and one can safely submerge electric motors and switch gear below its surface, the cryogenic temperatures contributing to low electrical resistance of copper and other metals.[31]

The physiological effects of LNG are closely linked to its cryogenic properties; while small spills of LNG on human or animal skin will rapidly evaporate, more extensive contact with LNG or thermally conducting materials at LNG temperature will cause damage to tissues such as frost bite, and in cases of prolonged exposure, embrittlement and complete destruction. Protective clothing worn by personnel exposed to potential contact with LNG must, therefore, be principally non-conducting and non-absorbent.[27]

Methane and other hydrocarbon vapours are non-toxic but can be asphyxiating. Ordinary gas masks with filters are therefore of no particular use in LNG installations, and complete self-contained breathing apparatus must be used to enter a contaminated area. Non-sparking tools, footwear and other equipment are, clearly, essential.

## 9.3 MATERIALS OF CONSTRUCTION AND EQUIPMENT FOR LOW TEMPERATURES

The cryogenic nature of LNG imposes certain restrictions on equipment and materials in contact with the liquid, and the safe handling of LNG demands strict adherence to suitable materials of construction.

In particular, mild steels undergo a drastic change in ductility at about $-50°C$, when they become extremely brittle, their Charpy V-notch impact strength dropping from about 30 to 5 ft lb or less. An alternative approach to the assessment of a suitable material for cryogenic temperature applications is the NDT (nil ductility transition) temperature, which should be well below $-160°C$ to ensure satisfactory performance over the range of temperatures which are likely to be encountered in LNG handling facilities.

Metals of suitably low NDT are nickel, copper, most aluminium alloys, many high nickel/chromium stainless steels and, among cheaper materials, certain 5% and most 9% nickel steels meeting ASTM specifications A-353-64, A-645-71 and A-553-65. Proposals to utilise alloy steels of even lower nickel content (3·5%) with other minor components, including molybdenum, for cryogenic applications so far apparently have not materialised. Composition and the more important mechanical properties of some of the alloys used in cryogenic equipment are shown in Table 9.1.[9]

The effect of low temperatures on mechanical and thermal properties of these materials of construction and their suitability for cryogenic duties are evident from Table 9.2 which lists Charpy V-notch toughness, lateral expansion, critical stress intensity, which relates to the material's ability to resist crack propagation, and thermal stress experienced during LNG loading, a function of expansion coefficient, elasticity and temperature drop.

An important aspect of LNG and cryogenic equipment is the large dimensional change which takes place when the equipment is cooled from ambient to cryogenic temperatures. Any design which does not allow for the differential expansion of metal shells and thermal insulation, storage tanks and their connections, pipes which are not used continuously, pumps, instruments and motors which combine several materials of different expansion coefficient, and similar

TABLE 9.1

Typical Metal Alloys Used for Cryogenic Duties—Composition and Mechanical Properties

| | Ferritic steels | | Austenitic steels | | Aluminium alloys |
| --- | --- | --- | --- | --- | --- |
| | Cryonic 5 | 9% Ni steel | SS304/L | Invar | Al 5083-0 |
| COMPOSITION (WT %) | | | | | |
| C | 0·08 | 0·08 | 0·03 | 0·07 | — |
| Mn | 0·50 | 0·65 | 1·50 | 0·44 | 0·40 |
| Ni | 5·00 | 9·00 | 9·00 | 37·00 | — |
| Si | 0·35 | 0·20 | 0·50 | 0·24 | 0·14 |
| Cr | — | — | 19·00 | — | 0·11 |
| Mo | 0·25 | — | — | — | — |
| Al | 0·08 | 0·03 | — | — | 95·20 |
| Mg | — | — | — | — | 3·90 |
| Fe | 93·74 | 90·04 | 69·97 | 62·25 | 0·25 |
| ASTM Code | A 645 | A 553/353 | A 240 | ASME Code | B 209 |
| Yield strength (MN/m²) | 447 | 586 517 | 173 | 242 | 117 |
| Tensile strength (MN/m²) | 655–790 | 690 | 483 | 447–552 | 269 |
| Percent elongation (min) | 20 | 20 | 45 | 30 | — |

equipment will result in mechanical failure after a few cooling and reheating cycles.

Safe practice, for instance, requires that double skinned tanks with expanded perlite insulation should be designed to withstand expansion or contraction of the inner shell without damage to the insulation, usually by providing an elastic blanket to take up movement of the inner shell and to prevent settling of the powder insulation. Pipes and connections entering through the shell have to allow for movement both of the pipe itself and of the wall through which it enters. Introducing all connections through the roof of the tank simplifies their design and obviates damage and breaks due to thermal expansion or contraction of either pipe or side wall.[3]

A design feature of metallic LNG tanks which may affect their safety is their ability to withstand the buoyancy of gaseous methane in the ullage space above the liquid. This produces an upward pull on the inner shell and, unless they are well anchored around their circumference, such tanks can fail owing to lifting of the side wall. Correct design of holding down bolts is therefore essential. Similarly frost heave, which can damage tank bottoms and result in tank

TABLE 9.2

Low Temperature Properties of Cryogenic Metal Alloys

| | Cryonic 5[a] | 9% Ni steel[b] | SS 304 L | Invar[c] | Al 5083-0 | ASTM requirement |
|---|---|---|---|---|---|---|
| Notch toughness (J at °C) | 61 at −171 | 41 at −196 | 163 at −196 | 68 at −196 | 30 at −196 | 27 at −171 27 at −196 |
| Lateral expansion (mm/mm × °C)[d] | 0·76 | 0·64 | n.a. | n.a. | n.a. | 0·38 |
| Critical stress intensity[d] (MPa/m at −171°C) | 387 | 298 | n.a. | n.a. | 155 | — |
| Thermal stress (when loading LNG)[e] (MN/m²) | 330 | 472 | 472 | 47 | 242 | — |

[a] Similar in composition and properties to UHB 2N50 (Uddeholm).
[b] 9% Nickel steels are made by US Steel, Uddeholm, etc.
[c] Invar is trade name of 36% nickel steel made by IMPHY Div. of Creusot Co.
[d] Measures tendency to propagate cracks at −171°C when load applied.
[e] Thermal stress when loading LNG, $S$, is proportional to modulus of elasticity, $E$, thermal expansion coefficient, $k$, and temperature drop $\Delta T$ equal to +21 to −161 = 182°C, i.e. $S = 182\,Ek$.

E

failure, must be prevented by insulation between tank and soil and, if necessary, by soil heating.

The cryogenic nature of LNG also exerts considerable influence on the design of heat exchangers. In particular, water used to heat and vaporise LNG must never be allowed to freeze. This can be achieved by use of an intermediate heat exchange fluid or by heating the water itself to well above freezing point or by ensuring a large enough flow to produce only a small temperature rise. Contraction of tubes or plates on cooling down to operating temperature must be allowed for and good thermal conductivity is essential to reduce temperature drop across the metal. Copper or aluminium meet these requirements and also have acceptable low temperature tensile strength and ductility.

Insulated cryogenic piping layout must be such that low temperature contraction does not result in pipe failure or separation from the insulation. Pipe material, as a rule, is aluminium, occasionally 9% nickel steel, and insulation polyurethane foam or polystyrene, protected against water penetration by an external coating of resin.[1]

Materials for valves, instruments and other fittings must be compatible with cryogenic temperatures, i.e. surfaces in contact with each other must have the same expansion coefficient. Copper, brass and stainless steel are used; rubbers and plastics must be avoided or carefully protected from contact with the cryogenic fluid or cold metal surfaces. Extended valve spindles ensure continuous insulation, but present problems of differential expansion.

Spills and leakages of LNG must be contained within bunding walls which are not affected by a cryogenic liquid. Compacted earth or prestressed reinforced concrete are commonly used.[8]

## 9.4 OFFICIAL STANDARDS FOR LNG FACILITIES

Experiences gained in the handling of LNG have been codified by a number of official bodies.

For instance, the National Fire Protection Association in its Standard No. 59 A[23] makes detailed recommendations regarding the design and operation of plant for the production, storage and handling of liquefied natural gas. Similarly the American Petroleum Institute has published code 2510 A[3] which describes the design and construction of LNG installations at petroleum terminals, natural gas processing plants, refineries and other industrial plants. Draft safety recommendations dealing with liquefied natural gas have also been published by the Institution of Gas Engineers in their communication 856,[17] and a similar standard is in preparation by the Institute of Petroleum, London. Local ordinances apply in several

US States, e.g. Massachusetts, New Jersey, New York and California.

Standards for the construction of above-ground metal tanks for the storage of LNG are summarised in Appendix Q of API Standard 620[4] which deals with materials, design, fabrication and construction of tanks for petroleum products.

Additional standards, mentioned and referenced in most of the basic standards, refer to concrete and concrete structures (American Concrete Institute), piping design and materials (ASME/API), vessels for cryogenic materials (ASME), ships, barges and other marine equipment (United States Coast Guard/NFPA) and refrigerants (NFPA). The ASTM have produced standards not only for the cryogenic metals already discussed but also for most other materials used in the construction of cryogenic installations.[27]

## 9.5 FIRE PREVENTION

The inflammability of air/methane blends downwind of an LNG spill results from the gradual diffusion of air and gas into each other.[24] Mixtures containing between about 5 and 15% methane are inflammable and their formation from LNG spills will depend mainly on wind speed and size of LNG pool, but also on ambient temperature, humidity, topography of the spill area if on land, etc. Experimental work has shown [16,29] that for a spill on water at a steady wind speed of 5 miles/hr an inflammable region will be formed 75 ft from the spill source for low rates of spill and as far out as 800 ft for a spill rate as high as 10 000 gal/min.

A mathematical model developed on the basis of 5-miles/hr winds was applied to an LNG pool 400-ft square; it showed that an inflammable mixture within 1400 ft from the pool edge existed 10 min after the occurrence of a spill. After 20 min the inflammable zone would spread to 9000 ft. Higher wind speeds, say 10 miles/hr, while tending to spread the inflammable cloud much further, also dilute the gas and result in a relatively thin shell of inflammable material, the methane concentration being rapidly reduced below the inflammable limit by turbulent mixing.

The methane concentration is invariably higher at the leading edge of the cloud, owing to rapid initial vaporisation of the LNG by contact with relatively warm containment dikes and bunds. As soon as heat transfer from surrounding structures is reduced by freezing or formation of gaseous layers evaporation rate appears to be much reduced; an obvious means of improving the efficiency of containment bunds is, therefore, the reduction of heat influx, e.g. by means of polyurethane insulation.

An important consideration in the design of LNG facilities is the

E*

eventual size of the LNG pool which would result from an LNG spill. Since rate of evaporation of the liquid is mainly a function of the surface area of the pool, there will be a maximum diameter, D, corresponding to a continuous spill rate, $m_s$, and an evaporation rate, $m_e$, which are related by the equation

$$D = 1 \cdot 13 \sqrt{(m_s/m_e)}$$

Both this correlation and the equation predicting the time taken for the maximum size pool to form

$$t = \sqrt{(m_s/4 \cdot 9\, m_e)}$$

strictly apply to LNG spilt on water, but can also be considered extremes of spread on land. Volumes spilt during the tests in question ranged from 10 to 10 000 gal and the size of the resulting pools varied between 6 and 200 ft.[29]

Fire fighting methods developed for LNG conflagrations aim more at control and containment of the fire than at complete extinction. Thus the main objective of fighting an LNG fire is to allow it to burn itself out without causing damage rather than to recover unburnt liquid.[6,10,31] The immediate objective must therefore be to cut off additional flow of liquid by means of strategically located shut-off valves or by transferring liquid from a damaged to an undamaged tank. At the same time adjacent tanks must be cooled externally by water spray; fog may also have to be used to ensure access to shut-off valves, drain covers, etc.

To extinguish an LNG fire due to a burst tank very large quantities of powder or foam would be required and it is therefore impractical to completely cover a diked impounding area or large tank surface; powder or foam extinguishers are only used to put out small fires and to deal with minor LNG leaks. Water spray and fog should be used to cool surrounding storage tanks and equipment and to confine the fire to the centre of the bunded area.[30]

The actual extinguishing of LNG fires, as distinct from their containment and the cooling of surrounding areas to prevent flame spread and reduce the effects of flame radiation, can be achieved by means of high expansion foams. Recent tests indicate that the effects of flame radiation from small burning pits (5–10 ft diameter) can be reduced to a safe level at distances equal to greater than one quarter the pool diameter. Optimum extinction characteristics are those of 500:1 expansion foams which are superior to the less dense (750 or 1000:1 expansion) foams in that they effectively insulate the residual liquid from the flame; the foam reduces the evaporation rate sufficiently and owing to the buoyancy of the burning gases can result in complete flame extinction. Where this is not the case a foam blanket at least reduces the downwind travel of ignitable vapour concentration near ground level, where the potential of flame spread is usually greatest. The presence of a foam blanket, as applied or after freezing, also provides support for a layer of dry chemicals for final extinguishment.[28]

## 9.6 THE INTERACTION BETWEEN LNG AND WATER

The danger of inflammable LNG mixtures is most acute if a cloud of gas vaporises from a pool of spilt liquid. The low temperature of the gas as well as the condensation of steam in the surrounding air will tend to confine such clouds and tend to prolong the existence of inflammable or explosive air/gas mixtures at the interface between gas cloud and air. Rapid dispersal of gas clouds is essential and, by the same token, possible sources of ignition or explosion must be kept away from LNG spills until all the liquid has evaporated and the cloud has dispersed.

It was therefore noted with some concern that LNG when spilt at sea or on any other water surface not only produced the expected inflammable vapour cloud which drifted downwind and eventually disintegrated, but also caused a series of explosions at or near the surface where liquid LNG was in direct contact with water.[22] The study of the non-chemical explosive interaction between LNG and water was therefore undertaken by a number of groups with a high degree of urgency.

Three theories were originally advanced to explain the occurrence of detonations when LNG was spilt:[10,13,18]

— the temperature difference between LNG and water is such that the liquid boils so violently that pockets of high pressure vapour are formed which expand suddenly and explosively;

— a shell of ice is formed and encapsulates droplets of LNG. When these warm up, evaporate and expand, the shell bursts producing an explosion;

— the Weber number, a dimensional group indicative of the rate of atomisation of a liquid, rises rapidly when a stream of vapour passes at high velocity through a pool of LNG; this results in even more rapid atomisation and very sudden expansion when warmer air or steam contacts the droplets. A shock wave and explosion could be the result.

In spite of the plausibility of these explanations recent theories appear to be at variance with all three postulated mechanisms. Work on liquid water in contact with molten metal, a system not too dissimilar to an LNG pool on the surface of the sea, indicates that fragmentation of the liquid occurs as a response to an external stimulus. The actual explosion takes place when the vapour envelope around the droplet collapses, in a manner not unlike the vaporisation of drops of water on a hot metal surface, first observed by Leyden-frost in the 17th century and known by his name.

Support for this theory comes from the observation that pure liquid methane rarely, if at all, interacts explosively with a water surface. LNG, which contains minor amounts of other hydrocarbons and certain impurities, will tend to detonate when spilt at sea, if the percentages of the various adventitious materials fall into a certain narrow range. Neither higher nor lower concentrations will result in explosions. It is therefore imaginable that the impurities present in liquid form—they will only evaporate at a much higher temperature —provide the stimulus for the onset of fragmentation. However, extensive further studies, some now under way, will be required before a complete answer to the problems of LNG/water explosions has been found.

## 9.7 STRATIFICATION OF LNG IN STORAGE AND ROLL-OVER

Since LNG is not made up entirely of liquefied methane but, in addition, may contain a number of other components, all the physical properties of the blend, and particularly its density and vapour pressure, can vary over a fairly wide range. However, density and vapour pressure of LNG are not only a function of its composition, but depend also on the temperature of the liquid. It is therefore imaginable that two LNG compositions, provided their liquid temperatures are different, could have exactly the same density. Similarly, one could visualise a colder liquid layer which originally had established itself at the bottom of an LNG tank to rise to the top as soon as thermal equilibrium had been established, and vice versa, for a warmer layer to move rapidly to the bottom as soon as thermal diffusion had evened out any temperature differences in the tank.[27]

Vaporisation rate of a liquid stored in a tank is usually a function of heat leakage inwards, and in the case of LNG boil-off can be partially selective, i.e. the lower boiling components can vaporise in preference to higher boiling materials. As a result, LNG stored over a period will change slightly in composition, the percentage of heavier hydrocarbons generally increasing in the course of time. Vaporisation will be rapid when a warm tank is filled and it is generally necessary to recompress any vapour formed during filling, unless some loss is acceptable and the gases which are generated can be vented or flared. In the absence of a vent, flare or compressor of sufficient capacity, vapour pressure in the tank will build up and eventually, if vaporisation turned out to be excessive, the tank would burst.

If there is incomplete mixing and the tank content consists of two or more layers of distinct composition, vaporisation from the lower layers can be suppressed or delayed by the hydrostatic pressure of the top layer or layers. It is therefore possible that very rapid vaporisation could take place if a supersaturated bottom layer, the vaporisation of which had been prevented by hydrostatic pressure, suddenly rose to the top as a result of the previously mentioned temperature equalisation within the tank. It is this phenomenon that has been described as 'roll-over' and that can result in an extremely dangerous situation if not countered or prevented.

An example of a narrowly averted accident due to the sudden roll-over of the content of an LNG storage tank is the following.

TABLE 9.3

Differences in LNG Composition Leading to Stratification

|  | Tank heel | Fresh cargo |
|---|---|---|
| Composition, mol % |  |  |
| Nitrogen | 0·35 | 0·02 |
| Methane | 63·62 | 62·26 |
| Ethane | 24·16 | 21·85 |
| Propane | 9·36 | 12·66 |
| n-Butane | 1·45 | 1·94 |
| isoButane | 0·90 | 1·20 |
| n-Pentane | 0·05 | 0·01 |
| isoPentanes | 0·11 | 0·06 |
| Density, lb/ft$^3$ | 33·82 | 34·06 |
| g/cm$^3$ | 0·542 | 0·546 |
| Pressures, mm H$_2$O, tank | 250 | 1 100 |
| bulk | 400 | 1 660 |
| Bubble point (calculated), °F | −249 | −245 |
| °C | −156 | −154 |

A 315 000 barrel tank (50 000 m$^3$) contained a 60 000 barrel heel of LNG. The analysis of this material, as shown in Table 9.3 and allowing for hydrostatic pressure, corresponded to a vapour pressure of 400 mm H$_2$O at the temperature of the heel and a density of 0·544 g/cm$^3$. Into this tank were rapidly transferred some 210 000 barrels of LNG from a ship which had been delayed for a period of about one month. As a result, the cargo had changed in composition, as shown in the second column of the table, and it was also denser and hotter than the residue in the storage tank.

The storage tank had a side-entering bottom nozzle and LNG was introduced in normal fashion. However, under the circumstances,

E**

little mixing occurred, as shown by the small amount of vapour generated during discharge, and eventually a potentially dangerous stratification with the less volatile, lighter, colder heel on top and the more volatile, heavier and hotter fresh cargo at the bottom resulted.

Over the 18 hr following unloading, the temperature of the original tank content gradually rose to that of the fresh cargo, with the result that rapid boil-off of methane occurred from the top layer. In consequence, the top layer increased in density and eventually became denser than the added material. As soon as this happened, it settled to the bottom, thus releasing the supersaturated vapour of the much larger volume of fresh material. The result, as one would expect, was a rapid evolution of methane vapour, copious release of gas through the vent to the flare in addition to vapour release through all safety valves, and a considerable pressure build-up inside the tank. Fortunately, the pressure could be contained and the tank did not burst. Emergency measures taken by the staff of the plant and the authorities—roads were closed and shipping alerted—prevented a major disaster.

A number of important deductions could be drawn from this experience. Clearly, stratification and roll-over could only occur if LNG cargoes of different composition were stored in the same tank, and by the same token stratification would not take place if the material consisted entirely of methane, i.e. there were no heavier components present. Clearly, neither bottom nozzles nor discharge into the vapour space of a tank provided a guarantee against the occurrence of stratification. In the tank in question very rapid charging through a mixing nozzle provided the answer,[25] but it would be difficult to arrange liquid mixing in ships' tanks, which are charged from the top through trunk-mounted injection nozzles.

An interesting theoretical consideration applies to the mixing phenomenon as such. Clearly, once convection currents are no longer confined to individual layers of uniform composition, mixing will take place and the dangers of rapid vapour formation will be eliminated. It appears now that small temperature differences, allied with relatively large differences in density, lead to convection currents confined to single layers and tend to prolong the period of stratification. Larger temperature differences between adjacent strata, on the other hand, seem to result in rapid mixing. If follows that introduction of a cargo which differs only slightly in temperature but considerably in density from the residue in the tank is particularly undesirable. Larger temperature differences, on the other hand, can be tolerated.

Attempts have been made to construct a model and to predict theoretically the rate of mixing and extent of delay of rapid vaporisation.[7] Making a number of basic assumptions in regard to diffusion rate and heat transfer between strata, it has been claimed that mixing and vaporisation rates can be predicted with reasonable accuracy.

In practice, a number of precautionary steps can be taken to prevent stratification and roll-over. It is generally sufficient to:
— prevent stratification by adequately mixing all incoming LNG with all liquid in the tank;
— limit the range of LNG compositions (and densities) added to the tank;
— prevent over-pressurisation of the tank by providing venting, flaring and recompression capacity sufficient to handle all the vapour generated during filling and subsequently.

However, the large size of the bigger LNG tanks makes thorough mixing difficult; to limit LNG composition may be possible if only one source of LNG is handled, and finally, roll-over in a very large tank may generate so much vapour that even the largest vents become limiting. A combination of all three approaches together with a limited capacity of the tank to withstand an internal pressure rise may, therefore, be the only practical approach to a solution.[32]

## REFERENCES

1. American Gas Association (1971). *LNG Safety Program*—Reports 1, 2 and 3, Arlington, Va. (1 and 2, A. D. Little, Battelle Memorial Inst. and University Engineers).
2. Anderson, R. P. and Armstrong, D. R. (1972). *Experimental Study of Vapour Explosions*, Session VI, Paper 3, LNG–3, Washington.
3. API 2510A, *Design and Construction of LNG Installations at Petroleum Terminals, Nat. Gas Processing Plants, Refineries and Other Industrial Plants*, 1968, Addendum 1969.
4. API Standard 620 Appendix Q (1968). *Materials, Design, Fabrication and Construction of Aboveground Metal Tanks for the Storage of LNG*.
5. Burgess, D. S. (1970). *Explosions from LNG*, US Bureau of Mines Report S-4105.
6. Burgess, D. S. and Zabetakis, M. G. (1962). *Fire and Explosion Hazards Associated with Liquefied Natural Gas*, US Bureau of Mines, Report R.I. 6099.
7. Chatterjee, N. and Geist, J. M. (1972). The effects of stratification on boil-off rates in LNG tanks, *Pipeline Gas J.*, **199**(11), 40–45.
8. Closner, J. J. (1971). Prestressed dike system for LNG storage, *Cryog. Ind. Gases*, **6**(5), 21.
9. Cordea, J. N., Frisby, D. L. and Kampschaefer, G. E. (1972). Steels for LNG storage, transportation, *Oil Gas J.*, **70**(41), 85–90.
10. Crouch, W. W. and Hillyer, J. C. (1972). What happens when LNG spills? *Chemtec*, **2**(4), 210–215.
11. Dyer, A. F. and Sommer, E. C. (1969). *Standards Development in the US for Liquefied Natural Gas Installations*, Proc. Int. Conf. LNG, pp. 138–148, London.

130     LIQUEFIED NATURAL GAS

12. Elliott, M. A., Seitel, C. W., Brown, F. W., Artz, R. T. and Berger, L. B. (1946). *Report on the Investigation of the Fire at the Liquefaction, Storage and Regasification Plant of the East Ohio Gas Co., Cleveland, Ohio, on Oct. 20, 1944,* US Bureau of Mines Report, R.I. 3867.
13. Enger, T. and Hartman, D. (1972). *Rapid Phase Transformation during LNG Installation on Water,* Session VI, Paper 2, LNG–3, Washington.
14. Frank, E. H. and Wardale, J. K. S. (1970). *Failure of a Liquid Ethylene Storage Tank,* Paper 6/7, LNG–2, Paris.
15. Gonzales, M. H., Subramanian, T. K. and Kao, R. L. (1968). *Physical Properties of Natural Gases at Cryogenic Conditions,* Paper 21, LNG–1, Chicago.
16. Humber-Basset, R. and Montet, A. (1972). *Flammable Mixtures Penetration in the Atmosphere from Spillage of LNG,* Session VI, Paper 4, LNG–3, Washington.
17. Inst. of Gas Engineers (1971). Safety Recommendations IGE/SE/II, Liquefied Natural Gas, Commun. 856.
18. Katz, D. L. and Sliepcevich, C. M. (1971). LNG/water explosions: cause and effect, *Hydrocarbon Proc.,* **50**(11), 240–44.
19. Klosek, J. and McKinley, C. (1968). *Densities of LNG and of Low Molecular Weight Hydrocarbons,* Paper 22, LNG–1, Chicago.
20. Maher, J. B. and van Gelder, L. R. (1972). Roll-over and thermal overfill in flat bottom LNG tanks, *Pipeline Gas J.,* **199**(11), 46–48, 113.
21. Massac, G. (1972). Safety of the sea transportation of LNG, *Tanker Bulk Carrier,* **18**(10), 14–16.
22. Nakanishi, E. and Reid, R. C. (1971). Liquid natural gas–water reactions, *Chem. Eng. Prog.,* **67**(12), 36–41.
23. National Fire Protection Association (1971). *Production Storage and Handling of Liquefied Natural Gas,* Standard No. 59A.
24. Parker, R. O. and Spata, J. K. (1968). *Downwind Travel of Vapours from Large Pools of Cryogenic Liquids,* Paper 24, LNG–1, Chicago.
25. Sarsten, J. A. (1972). LNG stratification and roll-over, *Pipeline Gas J.,* **199**(11), 37–39.
26. Stannard, J. H. (1971). New NFPA 59-A-1971 LNG Code, *Pipeline Gas J.,* **198**(10), 39.
27. Uhl, A. E., Amoroso, L. A. and Seiter, R. H., Safety and reliability of LNG facilities, *Gas,* **48**(11), 48–50, 1972; **48**(12), 34–36, 1972; **49**(1), 43–45, 1973; **49**(2), 36–38, 1973.
28. Welker, J. R., Wesson, H. R. and Brown, R. E. (1972). *Control of LNG Fires with High Expansion Foams,* ASME Paper 72-Pet-46, New Orleans.
29. Welker, J. R. (1972). *Vapour Dispersion from LNG Spills,* ASME Papers 72-Pet-48 and 72-Pet-61, New Orleans.
30. Walls, W. L. (1971). Fire protection for LNG plants, *Hydrocarbon Proc.,* **50**(9), 205–208.
31. Walls, W. L. (1972). LNG—a fire service appraisal, *Fire J.* **66**(1), 15–20.
32. Drake, E. M., Geist, J. M. and Smith, K. A. (1973). Prevent LNG 'rollover', *Hydrocarbon Proc.,* **52**(3), 87–90.

*Chapter 10*

# LNG and the Future

## 10.1 INTRODUCTION

The recent development of LNG technology, in regard both to handling and transport and to its applications, has been unusually rapid. Not only size and capacity of individual schemes have grown from relatively insignificant to enormous, but new materials of construction, novel types of equipment, unorthodox forms of storage, etc. have materialised over a period of ten to fifteen years.

It could be argued that such massive technical and commercial development permits one to extrapolate existing trends into the future with reasonable accuracy; in actual fact any attempt at forecasting LNG developments over the next twenty years or more is a risky exercise and may well result in substantial mistakes. Nevertheless it is proposed to try listing a number of technical advances and new applications and to assess their timing or the likelihood of their being realised. Additionally an attempt will be made at quantifying LNG manufacture and movements over the next twenty years, the maximum period over which it seems possible to foresee future construction.

## 10.2 FUTURE LNG TECHNOLOGY

Any assessment of future LNG technology can be broken down into the general areas of materials of construction, equipment and applications.

### 10.2.1 Materials of the Future

The main problem of LNG handling is the effect its low temperature exerts on the physical properties of the cheaper metals such as mild steel. In addition, the large temperature differences which occur in the course of transporting LNG present problems of thermal expansion. Materials of construction of the future should, therefore, be tough at temperatures down to $-170°C$ and have as low an expansion coefficient as possible. Clearly they should also be cheap, both as raw materials and in regards to processing.

There seems no doubt that both aluminium alloys and alloy steels, somewhat cheaper and also better than to-day's materials will be developed. As far as ferrous alloys are concerned the content of expensive alloying metals such as nickel and chromium may be reduced even further: after all Invar is made up of 37% or more of added metals and 8/18 stainless, 9% Ni steel and 5% Ni cryonic are all members of a progression which may well be capable of further extension.[1] There is no similar progression in aluminium alloys, but the price of 5088 aluminium has come down some 20% in monetary terms, i.e. much more if inflation is discounted, over the last few years. The content of metals other than aluminium is not significant and a change in composition would therefore not result in any price reductions.

Non-metallic cryogenic materials such as prestressed or post-stressed concrete and glass reinforced plastics are already used for LNG containment and transportation. Both have somewhat lower thermal expansion coefficients than have metals and do not suffer from low temperature embrittlement. There is very little doubt, therefore, that their use will expand very substantially. One particular application, ship construction, seems to call for further improvements, mainly because of the danger of damage to ships' hulls if, owing to interior leakage, LNG reaches the outer mild steel plates. Construction of ships in concrete, so far confined to very much smaller vessels, may well be the answer.

Improved forms of insulation could undoubtedly reduce the cost of cryogenic plant construction. Unfortunately it seems unlikely that revolutionary changes will take place in this area. While a vacuum or low pressure gas minimises heat transfer the construction of vacuum insulated vessels must remain expensive and the resulting double walled vessel will always be relatively fragile. Replacement of the vacuum by stagnant gas—and this, essentially, is what insulating materials do—reduces heat transfer to low values. But the differences in heat transfer rate between the various forms of lagging are relatively insignificant.

There are, of course, differences in structural properties, and some materials such as polystyrene and polyurethane are self-supporting, while others such as perlite, are not. It is believed that improvements in the structural strength of insulating materials will be made over the next few years, but substantial reductions in heat transfer rate are considered unlikely. The extended use of complex or sandwich materials to ensure both thermal insulation and liquid containment, which are prefabricated in suitable shapes and joined on the site of the LNG plant or shipyard, is another likely improvement, in line with construction techniques developed in the recent past.

## 10.2.2 Equipment of the Future

There are numerous items of equipment ranging from valves to tanker loading arms, from gas compressors to specially designed roofs for in-ground tankage, from alloy containment membranes to submerged cryogenic pumps, which have been developed over the last few years to meet the specific needs of LNG plants. There is very little doubt that similar developments will take place over the years to come, although it is difficult to predict whether they will be equally revolutionary in character as the first batch of improvements listed.

The design of compressors and expanders for natural gas liquefaction plants is undoubtedly due for further improvement. Number of stages, compression ratios, combinations of centrifugal and axial compression and of turbo-expander and expander valve cooling in accordance with the composition and characteristics of the refrigerant which is being cooled, are becoming a matter of detailed thermodynamic calculation rather than rule of thumb selection.[3]

Similarly heat exchanger design is no longer based on experience but temperature difference, pressure drop, extent of liquefaction, form of boiling, i.e. nucleate or film, are calculated and optimised by means of computer techniques. While no complete liquefaction plant has so far been designed entirely on the basis of computed compressor input, heat exchange and expander performance, considerable progress has already been made towards minimising exchanger temperature differences in order to approach reversibility and optimum efficiency of the plant while operating within economic limits. Further progress in computerised design of liquefaction plants tailored to specific natural gas compositions is therefore considered certain.

Developments in heat exchanger design have concentrated on spirally wound and, more recently, on plate-fin exchangers. Neither seems to have reached the end of the line and, particularly, manufacturing techniques for plate-fin type cores could be further improved. The present method of immersing aluminium plates in a solder or brazing alloy of a melting point just below that of the corrugated plates depends on uniform temperatures to ensure an even spread of the solder; very large exchanger cores cannot be evenly preheated and this technique, which could substantially reduce the cost of cryogenic heat exchange, is restricted to the smaller sizes at the moment. There is little doubt that novel methods of heating, e.g. fluidised beds, will improve this situation and permit the construction of larger units. Alternatively, lower melting solders of acceptable mechanical properties could produce the same ultimate effect of cheaper cryogenic heat exchange.

The conditions under which cryogenic tanks are assembled are not normally conducive to accurate workmanship and precise construction. It is difficult, for example, to weld containment membranes or to join membrane/insulation sandwiches in a shipyard or a tankage site to the high standards required to prevent localised leaks or insulation discontinuities. Methods of construction which permit workshop assembly of entire ships' tanks or of large sections of storage vessels have consequently been developed—a trend which has favoured the use of self-supporting tanks in ships and barges at the expense of the basically cheaper membrane designs. Advances are expected to take place in two areas: firstly, self-supporting tanks, specifically designed for structural rigidity—the Kvaerner Moss spherical tanks are an instance—should be further improved to make optimum use of hull space, and secondly, improved primary and secondary membranes to protect mild steel hulls against LNG penetration will undoubtedly evolve.

The design principles of vaporisers for the reconversion of LNG into a gaseous fuel are relatively simple. A cheap source of heat such as ambient air, river or sea water is used to heat the external surfaces of a pressurised heat exchanger, the boiling liquid passing through the interior of pipes, coils or vessels with extended surfaces. Improved heat flux, particularly between metal and LNG, can be achieved by nucleate rather than film boiling, and surface coatings to achieve this have recently become available.[7] Their widespread use in all forms of heat exchange which involves boiling cryogenic fluids, i.e. in LNG revaporisers, condensing exchangers in liquefaction plant, etc. is expected. Similarly, the use of porous-surface coatings will no doubt become more widespread in future.[2,5]

Improvements in instrumentation for the measurement of volume, flow, density, temperature and calorific value of LNG are becoming of considerable importance. In line with refinery and oil industry instrumentation trends the replacement of measured by computed values obtained from composition and temperature only, is making considerable progress. Automated chromatographs, for example, can provide as much information as density meters, calorimeters, Wobbe meters, viscosity meters, etc. and operate independently of temperature measurement. Problems of volume and flow measurement at cryogenic temperatures will also be reduced in severity by computation based on accurate physical data at low temperatures for both containing vessel and liquefied gas. The great importance of determining accurate thermodynamic data and physical properties of hydrocarbon blends at low temperatures follows logically from these considerations. Studies in these fields are already under way and are also planned for the future.[4]

## 10.3 FUTURE LNG APPLICATIONS

Relatively few of the LNG applications mentioned in Chapter 8 have so far materialised; automotive use is confined to experimental units, and suggested aviation applications are the outcome of calculations and economic evaluation rather than experiment. Similarly, the advantages obtainable from the use of LNG for refrigeration are expected rather than actual. The aim of the present discussion is therefore mainly to assess the likelihood and timing of their realisation.

### 10.3.1 Fuel Applications

The great advantage of LNG over other fuels is its ability to burn without the emission of impurities such as sulphur dioxide, soot, lead oxide and part-burnt hydrocarbons. Furthermore, whenever LNG is carried over a distance part of it will evaporate through heat leakage; a second advantage for LNG burning is therefore the fact that vapour consumed during transport need not be recompressed or lost.

The main drawback of any LNG burning system which relies entirely on natural vaporisation from storage tanks is the lack of balance between heat leakage and consumption by the engine. Since the vapour, once it is formed, must be consumed or lost, any such system becomes extremely inflexible. Additional freedom can be provided by means of a vaporiser which will supply any shortfall of fuel, but to deal with the opposite situation, i.e. excess fuel vaporisation, expensive vapour recompression facilities are required, or at least a reduction of heat leakage and natural vaporisation to the absolute minimum must be achieved.

The design of improved insulated storage tanks for road vehicles and aircraft thus becomes of considerable urgency. In fact it has been claimed that automotive fuel tanks from which less than $0.1\%$ of the LNG vaporises in 24 hr have been designed. It is doubtful, however, whether even such a rate is acceptable from an economic and safety angle if, for example, a vehicle is garaged over-night or over a week-end. Further improvements in storage tank design and insulation are under study and results are expected to become available shortly.

The really widespread use of LNG as an automotive fuel is further inhibited by its limited availability; in consequence vehicles, unless they are used purely locally, must have dual fuel capability which adds substantially to cost and space occupied by fuel tanks. Desirable automotive LNG improvements, consequently, would be very low cost conversion kits and use of the same tank for both fuels. Little or no developments in this area are believed to be in progress.

The development of aviation and particularly automotive use is inhibited by the knowledge that overall availability of LNG will be limited; it would certainly be impossible to convert more than a small percentage of the world's motor cars to LNG, and even substitution of a large proportion of jet fuels used at present would soon exhaust available LNG supplies. Resorting to LNG in particular circumstances e.g. by vehicles in the centres of large cities, by aircraft during take-off from congested airports and similar uses, would therefore seem appropriate. In fact the use of insulated LNG containers which could be folded or jettisoned after use could provide an answer to air pollution by full throttle operation of aircraft during take-off. Development of appropriate containers should be given serious consideration.

Similarly, from a purely environmental angle, cities with a high incidence of air pollution should aim at converting public transport, taxis and locally operating goods vehicles to LNG, rather than consider unlimited sales of LNG as a motor fuel. Where helicopter traffic between airports and the city centre is heavy, conversion of these and other vertical take-off machines should be given priority, bearing in mind that helicopters are generally weight- rather than volume-limited. Replacement of gasoline and diesel fuels in marine engines for boats using pollution sensitive waters such as lakes, harbours and small rivers would seem highly desirable. Finally, diesel locomotives while stationary or pulling out of stations in cities could reduce air pollution by switching to LNG.

Developments in most of these areas could be accelerated by a clearcut anti-pollution stance on the part of national governments and local administrations. Legal enforcement would, of course, result in rapid penetration of these markets. However, even pre-ferential tax treatment of clean fuels could provide the necessary encouragement, bearing in mind the extra cost of conversion and insulated tankage, and the relatively high cost of the fuel itself.

### 10.3.2 Low Temperature Applications
None of the many LNG projects which have recently been completed in the US, Japan and Europe is so far using the heat capacity of the vaporising liquid as a source of industrial refrigeration. The reasons for such seeming profligacy are several: refrigeration is rarely required in the same location and it is difficult to transfer the cold elsewhere; proposals to move existing plant requiring cold in large

quantities or to construct new facilities are, as a rule, rejected because of fluctuating rates of LNG use and therefore of the cold available. Full use of available refrigeration at all times in a continuously operating plant would thus require stand-by refrigeration facilities, the cost of which would balance out any savings due to cheap LNG derived cold.

The obvious advance required in this field would be storage of excess refrigeration which could be used during periods of low gas use. Unfortunately there is little prospect of such a system being developed, refrigeration being normally stored in the form of latent heat of the denser phase and released by a phase transition, much as in natural gas liquefaction and revaporisation. It seems unlikely that a more compact system operating at or near the boiling point of LNG will be discovered.

The alternative to storing refrigeration is to use the cold in a 'seasonal' application. Again the availability of more cold during periods of intensive gas demand is in conflict with more demand for air conditioning, refrigeration of foodstuffs, and other uses of cold during the summer months.

Most potential users of LNG cold are, therefore, in practice restricted to base load LNG plants, and slaughter houses, air separation plants, frozen food packaging and similar activities will, no doubt, grow up in and around such facilities.

In such situations it is important to transfer heat, i.e. negative refrigeration, efficiently from the materials which are to be cooled to the heat sink, the stored LNG. Novel methods of heat transfer over a distance will, no doubt, play their part, and the recent development of efficient heat pipes should contribute to the successful use of LNG refrigeration. The principle of the heat pipe, flow of vapour in one direction, condensation of the vapour and its return as a liquid through a capillary wick, has been known for some time to be unusually efficient but has so far found few applications, and none in refrigeration.

The only industry whose activities seem to run parallel with normal LNG demand is electric power generation and transmission. Use of LNG refrigeration to attain super-conductivity in certain metals, particularly copper, therefore seems logical, and a number of projects are reported to be under consideration.[5] Unfortunately, to attain genuine super-conductivity a temperature some 120°C lower than that of LNG is required and the use of liquid helium rather than LNG is essential. LNG can, however, be used to precool helium before liquefaction of the latter, and a proposal to transport LNG and electric power in a combined pipeline/cable with additional cooling by means of a liquid helium loop has been evaluated and claimed to be economically attractive for the joint transport of electricity and LNG.[8] Applications, apart from distribution of electricity and gas from generating station and adjacent LNG plant, could be found in long distance transmission of electric power to a gas field while piping LNG in the opposite direction, e.g. in Northern Canada, Alaska or Siberia. Other proposals under consideration are the generation of electricity in alternators submerged in cryogenic fluids to reduce their internal resistance, or the design of refrigerated, low resistance electromagnets to improve the generating efficiency of these plants.

Proposals to combine refrigeration with power generation, e.g. in Stirling engines, thermoelectric generators and similar power sources requiring both a heat source and a cold sink, are handicapped by the inherent lack of balance between the latent heat of vaporisation of a hydrocarbon and its heat of combustion. In other words it takes some 211 kcal to vaporise one kilogram of methane and raise it to room temperature; however, when oxidised to carbon dioxide and water the same mass of LNG will release 13 250 kcal. While the efficiency of a thermopile, the cold junctions of which are refrigerated by means of vaporising LNG, would thus be slightly better than that of an air or water cooled unit, the improvement at present does not appear sufficient to outweigh the other drawbacks of thermoelectric generators (D.C., low voltage, large bulk and inherently low efficiency).

Similar considerations apply to the use of LNG in refrigerated transport: unless insulation is unusually good refrigeration requirements will far exceed the available 1·6% of combustion energy, and additional cooling over and above that available from fuel vaporisation will be needed.

## 10.4 MISCELLANEOUS DEVELOPMENTS

Apart from improvements in the technology of LNG production, handling and applications we would expect a better understanding to develop of LNG properties which have a bearing on safety such as the phenomenon of the explosive non-chemical interaction with water, and the prediction of inflammable gas mixtures downwind of an LNG leak or discharge.

It is appreciated that the chemical and physical processes involved in any discharge of a cryogenic fluid on land, on water or in the air are complex: the presence of water vapour and impurities in air, of liquid water and air in soil, rock or bund wall surrounding an LNG spill and the impurities in sea water, for example, can all exert some effect on revaporisation rates, formation of inflammable mixtures and ignition mechanisms. Furthermore, the composition of the liquefied gas itself appears to be critical with regard to the only partly understood small explosions which occur when LNG is spilt on water, and must also have some bearing on the formation of inflammable clouds downwind of LNG leaks or spills. Nevertheless there are hopes of a better understanding of the mechanisms involved through experimental work which is now in progress.[6] It seems reasonable to expect improved design of safety features such as retaining dikes, safety valves and flares, collecting sumps and cut-off valves. A better appreciation of and more logical codes for safety distances for all types of LNG installations should follow.

Another as yet unsolved problem is the long distance transport of LNG by insulated pipeline. Problems of materials of construction, optimum operating pressure, type of insulation and calculation of pumping and cooling requirements, particularly under conditions of varying throughput, i.e. from initial cooldown via various operating rates to occasional shut-down, are at present under study.[9] Improved mathematical models will, no doubt, be developed and will be used to calculate optimum design at minimal investment and operating costs. It has in fact been claimed that already on the basis of present technology transportation of natural gas over a distance of 1000 miles at the rate of $10^9$ cfd costs about the same whether the gas is liquefied or in the gaseous state. Replacement of long distance high pressure natural gas trunk lines by insulated LNG lines is therefore ultimately imaginable.

An interesting modified form of liquefied gas which may find uses in the not too distant future is so-called 'slush' LNG. Methane in its frozen form with a certain amount of liquid adhering to the solid is not only somewhat denser than LNG (about 10%) but is easier to transfer from one vessel into another and when transported is less inclined to surge during acceleration or deceleration. Construction of storage tanks and transfer equipment are thus simplified and slush LNG may well become the preferred fuel for aircraft, automobiles and other transport equipment and may also become the form in which LNG is transported by pipeline.

A somewhat different approach to natural gas compression and transportation has been employed by Columbia Gas who claim advantages for their so-called CNG (compressed natural gas) and MLG (medium condition natural gas). Their argument is that compression and storage costs can be reduced by operating at about 75–80 atm pressure and $-60°C$, or better still at only 13–14 atm and $-115°C$.

This eliminates the need for a full scale liquefaction plant, and in the case of CNG anyway, low cost materials of construction become

140 LIQUEFIED NATURAL GAS

acceptable. However, the cost of high pressure containers, particularly of aluminium cargo bottles for shipboard use which would withstand this high pressure, has induced Columbia to abandon CNG and to adopt the MLG route.

MLG is said to require only half the horsepower of an analogous LNG plant. Larger diameter cylinders compared with those required for CNG were to be used for on-board storage and in its latest version MLG on land was to be stored in prestressed, insulated concrete tanks. Clearly one could also use depleted gas wells, salt cavities and other forms of under-ground or in-ground storage.

Although no MLG facilities have so far been constructed its developers claim that it is ready for commerical use.[10]

## REFERENCES

1. Cordea, J. N., Frisby, D. L. and Kampschaefer, G. E. (1972). Steels for LNG storage, transportation, *Oil Gas J.*, **70**(41), 85–90.
2. Crawford, D. B. and Eschenbrenner, G. P. (1972). Heat transfer equipment for LNG projects, *Chem. Eng. Prog.*, **68**(9), 62–70.
3. Hallock, D. C., Farber, R. M. and Davis, C. C. (1972). Compressors and drivers for LNG plants, *Chem. Eng. Prog.*, **68**(9), 77–82.
4. Lee, B. I., Erbar, J. H. and Edminster, W. C. (1972). LNG—thermodynamic properties at low temperatures, *Chem. Eng. Prog.*, **68**(9), 83/4.
5. Milton, R. M. and Gottzman, C. F. (1972). High efficiency reboilers and condensers, *Chem. Eng. Prog.*, **68**(9), 56–61.
6. Nakanishi, E. and Reid, R. C. (1971). Liquid natural gas/water reactions, *Chem. Eng. Prog.*, **67**(12), 36–41.
7. O'Neill, P. S. and Terbot, J. W. (1972). *Impact of New Heat Exchanger Technology on Large LNG Plants*, Session II, Paper 4, LNG–3, Washington.
8. Pastuhov, A. and Ruccia, F. (1970). Why not transport LNG and electricity at the same time? *Pipe Line Ind.*, **32**(5), 44–48.
9. Walker, G., Coulter, D. M. and Narrie, D. H. (1972). *Long Distance Natural Gas Pipelines*, Inst. Gas Engs. Communication 872, London.
10. Broeker, R. J. (1972). *CNG and NLG—New Natural Gas Transportation Processes*, LNG Economics & Technology, Energy Communications Inc., pp. 138–40, Dallas.

# Thermodynamics of Refrigeration

The liquefaction of a gas requires the production and maintenance of very low temperatures. A mechanism is needed to absorb heat at the low temperature of the refrigerated system and to dissipate heat at or above the temperature of the environment. Equipment to carry out such an energy transfer is referred to as a 'heat pump' and the liquid or gas which is generally needed to operate the process is known as the working fluid.

A heat pump or refrigerator then operates between two given temperatures and uses a certain amount of mechanical energy to effect the heat transfer from the lower to the higher temperature. Optimum design of the mechanism and a careful selection of the working fluid should ensure the highest efficiency, i.e. the maximum amount of refrigeration for the lowest expenditure of mechanical energy.

Since any heat or other energy transfer which is not reversible must be larger than the corresponding reversible change, minimum energy consumption will clearly be associated with a reversible process. Furthermore, in order to ensure full utilisation of heat and energy the most efficient heat pump will operate in cyclic fashion, i.e. after completion of the energy transfer the working fluid will be returned to its original condition.

The second law of thermodynamics can be formulated as

$$\mathrm{d}Q \leqq T\mathrm{d}S \tag{1}$$

and

$$\mathrm{d}Q_{\mathrm{rev}} = T\mathrm{d}S$$

for a reversible change where $\mathrm{d}Q$ is the element of heat absorbed by a system, $T$ is the absolute temperature and $\mathrm{d}S$ the entropy change.

Similarly the first law states that for any energy change within a closed system

$$\mathrm{d}Q = \mathrm{d}E + \sum \mathrm{d}w \tag{2}$$

where $\mathrm{d}E$ is the change in intrinsic energy and $\mathrm{d}w$ the work done by the system. The latter can, of course, be of many different types, i.e. mechanical, magnetic, electrical. But invariably it is made up of two elements of which one is in the nature of a displacement, strain or change in an external variable $(X)$ and the other is a force $(Y)$. It follows that

$$\mathrm{d}w = \sum Y_i \, \mathrm{d}X_i \qquad (3)$$

and

$$\mathrm{d}Q_{\mathrm{rev}} = \mathrm{d}E + \sum Y_i \, \mathrm{d}X_i = T \, \mathrm{d}S \qquad (4)$$

or for any reversible process

$$\mathrm{d}E = T \, \mathrm{d}S - \sum Y_i \, \mathrm{d}X_i \qquad (5)$$

Introducing next a function $F$, the Gibbs free energy of the system, and defining it as

$$F = E - TS + \sum Y_i \, \mathrm{d}X_i$$

or

$$F = H - TS \qquad (6)$$

where $H$ is the enthalpy of the system, it follows that

$$\mathrm{d}F = \mathrm{d}E - T \, \mathrm{d}S - S \, \mathrm{d}T + \sum Y_i \, \mathrm{d}X_i + \sum X_i \, \mathrm{d}Y_i$$
$$\mathrm{d}F = -S \, \mathrm{d}T + \sum X_i \, \mathrm{d}Y_i \qquad (7)$$

Among the types of energy which can be used in a heat pump to transfer heat from a lower to a higher temperature a fairly common form is the thermal expansion of the working fluid producing either a pressure change at constant volume or a volume change at constant pressure. In systems of this type

$$\sum X_i \, \mathrm{d}Y_i = P\mathrm{d}V$$

and

$$\mathrm{d}F = -S \, \mathrm{d}T + V \, \mathrm{d}P \qquad (8)$$

or

$$\left(\frac{\partial S}{\partial P}\right)_T = -\left(\frac{\partial V}{\partial T}\right)_P \qquad (9)$$

and the heat absorbed by reversible isothermal compression of a gas from $P_1$ to $P_2$

$$\Delta Q = T\int_{P_1}^{P_2} \left(\frac{\partial S}{\partial P}\right) \, \mathrm{d}P = -T\int_{P_1}^{P_2} \left(\frac{\partial V}{\partial T}\right)_P \, \mathrm{d}P$$

is obtained by combining eqns. (4) and (9).

The relationship $(\partial V)/(\partial T)_P$ is obtainable from the equation of state of the working fluid, and clearly for a significant transfer of heat to take place it will be necessary to use a working fluid which undergoes a substantial change in volume over a reasonable temperature range, in other words a gas. If the gas is ideal, i.e.

$$PV = RT$$

the isothermal heat released on compression from $P_1$ to $P_2$ will be

$$\Delta Q = -T \int_{P_1}^{P_2} \left(\frac{\partial V}{\partial T}\right)_P \mathrm{d}P$$

$$= -T \int_{P_1}^{P_2} \frac{\mathrm{d}P}{P} R$$

$$= -RT \ln \frac{P_2}{P_1} \tag{10}$$

If the system undergoes a phase change (as a rule vaporisation), i.e. if both liquid and vapour are present in equilibrium, one can no longer think in terms of a gas expanding at constant pressure, since this would only occur in a two-phase system at constant temperature. The external work done by such a system is due to the change in specific volume, and the heat absorbed in an isothermal expansion

$$\Delta Q = T \int_{V_1}^{V_2} \left(\frac{\partial S}{\partial V}\right)_T \mathrm{d}V \tag{11}$$

where $\Delta Q = L$, the latent heat of vaporisation, and $V_1$ and $V_2$ are the specific volumes of liquid and vapour respectively.

It is of interest to compare typical values for the heat transfers represented by eqns. (10) and (11). Assuming for instance a compression ratio of ten and operation at room temperature (300 K), a perfect gas with a molecular weight of 44 (propane) will absorb 32·3 cal/g. A two-phase system of the same gas in equilibrium with its condensate will have a latent heat of vaporisation of about 127 cal/g, i.e. much more heat will be transferred by the same mass if a phase change is involved.

The empirical equation for the latent heat of vaporisation

$$L = \frac{20T - 307}{M} \text{ cal/g}$$

where $M$ is the molecular weight and $T$ the absolute temperature, indicates that optimum heat transfer per mass of working fluid is obtained with low molecular weight fluids, provided of course they can be condensed at reasonable pressures. Ammonia, the lower fluorocarbons (Freon), propane and water are therefore suitable working fluids at room temperature. At lower temperatures ethylene, and at even lower temperatures, methane become technically acceptable.

Having chosen a suitable working fluid for refrigeration it next becomes necessary to devise an appropriate working cycle which will absorb heat at low temperature, reject it at a higher temperature and restore the working fluid to its original condition. Ideally all system changes involved in such a cycle should be reversible and a number of reversible thermodynamic cycles have been thought of which would meet these requirements.

The *Carnot* cycle, the first of the refrigeration cycles analysed by physicists, consists of two isothermal heat absorption and rejection steps at two different temperatures, and two adiabatic (or isentropic)

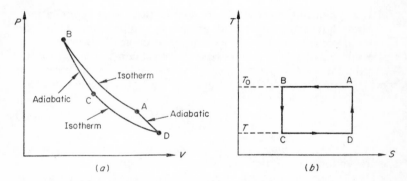

FIG. A.1    Carnot cycle.

changes connecting the extremes of the isothermal changes. On a pressure/volume diagram (Fig. A.1a) for an ideal gas, where adiabatic changes have steeper and isothermal changes have flatter slopes, line AB represents the isothermal rejection of heat at the higher temperature $T_0$, BC the adiabatic expansion resulting in a temperature drop to $T$, CD the isothermal absorption of heat at $T$ and DA the adiabatic compression of the gas, which results in a temperature rise to $T_0$. On the corresponding temperature/entropy diagram (Fig. A.1b) letters and lines have the same significance and the Carnot cycle is represented by a rectangle. In both diagrams the enclosed area represents the work done on the system to produce the indicated heat transfer.

The nearest approximation to the Carnot cycle would be a refrigerator consisting of two 'isothermal' heat exchangers, an adiabatic gas compressor and an adiabatic expansion engine, the compressed gas being 'cooled' at the temperature of the heat sink, expanded without heat absorption through the engine, and 'heated' isothermally at the refrigeration temperature.

The *Stirling* refrigeration cycle consists again of two phases of isothermal heat absorption and rejection respectively. However, the cycle is completed by two constant volume pressure changes which are no longer adiabatic, as shown in $P/V$ and $T/S$ form in Figs. A.2a and A.2b. Line AB again represents the isothermal dissipation of heat while the gas is compressed. Pressure is now reduced but the volume of the gas is kept constant, and this results in a temperature drop from $T_0$ to $T$ (line BC). This is followed by isothermal expansion (CD) with heat absorption at temperature $T$. A constant volume compression with heating of the gas to $T_0$ completes the cycle (DA).

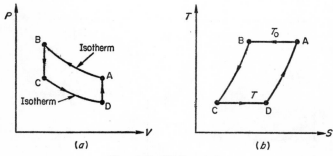

FIG. A.2    Stirling cycle.

It will be noted that in the Stirling cycle heat is rejected during the constant volume (isochoric) pressure reduction BC and the same amount of heat is absorbed during compression DA. The simplest method of approximating such a cycle is, therefore, heat exchange between the working fluid under compression and the working fluid under expansion. A Stirling cycle heat engine thus consists of two 'isothermal' compressors which take their suction from and discharge into a regenerative heat exchanger. By operating the two pistons 90° out of phase, i.e. the pistons are coupled such that part of the compression is isochoric and part isothermal, it becomes possible to transfer heat from the lower to the higher temperature.

A variant of the Stirling cycle is the *Ericsson* cycle in which isothermal compression AB and expansion CD are linked by constant pressure (isobaric) volume changes BC and DA, as shown in Figs. A.3a and A.3b. In theory heat transfer between the two isobaric phases could be by counterflow heat exchange, and again the two pistons effecting isothermal expansion and compression should be 90° out of phase and linked.

The *Claude* cycle, also known as the Siemens cycle, can be considered a variant of the Ericsson cycle in which the isothermal expansion phase is replaced by an adiabatic pressure/volume change.

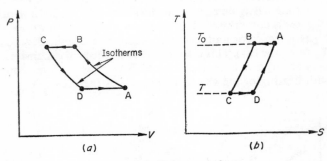

FIG. A.3    Ericsson cycle.

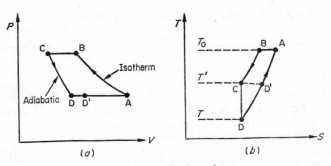

FIG. A.4    Claude cycle.

This results in a greater refrigeration effect as shown in Fig. A.4b, but implies that heat will be absorbed over a range of temperatures from $T$ to $T_0$, rather than at one temperature only.

The cycle consists of isothermal compression (AB), heat exchange at constant pressure between the hot and cold streams (BC and DA), adiabatic expansion with cooling to $T$ (CD), followed by isentropic expansion with heat absorption from the refrigerator (DD) and from the compressed working fluid (DA).

Expansion of a gas through an orifice rather than through an expansion engine seems at first sight very much less efficient. However, returning to eqn. (2) in the form of

$$dQ = dE + P\,dV$$

where $dE$ represents the internal and $P\,dV$ the external work done on a system, we find that merely throttling the flow of a fluid with a resultant pressure and volume change across an orifice represented by the equation

$$\Delta Q = \Delta E + P_1 V_1 - P_2 V_2 = H_1 - H_2$$

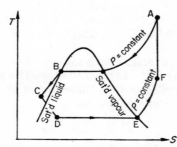

FIG. A.5   Vapour compression refrigeration cycle.

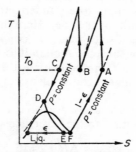

FIG. A.6   Joule–Thompson refrigeration cycle (two-stage compression with intercooling).

results in substantial changes in internal energy, whereas the external term $P_1 V_1 - P_2 V_2$ is very small in the absence of a phase change and fairly small even where a saturated liquid is converted into a superheated vapour. It is therefore not essential, particularly where the gas does not condense when passing through the orifice, to recover the external work released in throttling a gas. In fact there is no systematic difference in efficiency between expander refrigeration systems (Claude cycle) and Joule–Thompson cycles (single or cascade) or Stirling cycles, although the first and last operate on the expansion engine and the second on the expansion valve principle.

The difference between refrigeration cycles involving condensation and revaporisation on the one hand (vapour recompression cycles) and simple Joule–Thompson cycles on the other is shown diagramatically in Figs. A.5 and A.6.

In the former the working fluid must be below its critical temperature and above its critical pressure when leaving the first heat exchanger (point C). In Joule–Thompson refrigerators, on the other hand, the fluid is injected into the heat exchanger well above its critical temperature (point D) and it may take several complete cycles before it is refrigerated sufficiently to produce liquid on being expanded.

A further point to bear in mind is the sign of $(\partial E)/(\partial V)_P$ which can be positive, negative or zero. In practice Joule–Thompson refrigerators operate over a range from three times up to six times the critical temperature of the working fluid, i.e.

$$6T_c > T_2 > 3T_c$$

since at temperatures higher than six times critical heat is liberated rather than absorbed when a gas is expanded through an orifice or throttle valve, i.e. $(\partial E)/(\partial V)_P$ is negative.

No such restrictions apply to expander refrigeration cycles using expansion engines (piston or turbine). Here the condition of the gas after expansion is critical since one does not wish to produce liquid in the expander itself in substantial quantities. The gas should therefore not be cooled below its dew point at the turbine exhaust pressure; a final expansion through a throttle valve completes liquefaction in a Claude cycle.

Actual refrigeration processes are of course made up of non-reversible steps. In addition to the reversible work

$$W_{rev} = \Delta H + T_2 \Delta S \tag{12}$$

where $\Delta H$ is the enthalpy difference between the two states of the gas, $\Delta S$ the entropy difference and $T_2$ the average temperature of heat dispersal, real refrigeration absorbs additional energy represented by

$$W_{irrev} = W_{rev} + \sum T \Delta S_{irrev} \tag{13}$$

where $\Delta S$ is the entropy increment of each irreversible step and $T$ the average temperature.

Irreversibility in refrigeration cycles is basically due to three factors:

— the need to maintain a finite temperature difference across heat exchanger surfaces and between gases and regenerators;
— loss of heat to and gain of heat from the atmosphere due to imperfect insulation;
— mechanical losses in compression and expander equipment due to metal friction and in pipes due to fluid friction.

The efficiency of large gas liquefaction plants can be expressed as the ratio of theoretical (reversible) work required to convert the gas at ambient temperature and plant inlet pressure into a liquid at atmospheric pressure, divided by the actual energy consumption of the plant. Both reversible and actual energy requirements will be affected by the initial condition of the gas (pressure and temperature) and the temperature of the heat sink (average cooling water or ambient temperature). To compare different cycles it is therefore essential that allowance be made for differences in the above variables.

F

## Appendix B

# Approximate Economics of an LNG Scheme

(All figures are orders of magnitude only)

The purpose of this calculation is to estimate the gas price increase which would be required to meet the cost of substituting 25% or 50% of the send-out volume of a utility company by imported LNG. A number of assumptions must be made, among them the following.

Natural gas is purchased at present out of a pipeline at an average price of 50 ¢/MM Btu. The quality of the gas, as distributed at present, is such that the introduction of revaporised LNG will not produce interchangeability problems. Similarly there will be no major changes in distribution facilities other than those already allowed for in the LNG investment.

The scheme itself, it is assumed, will consist of a liquefaction plant, LNG storage at both loading and delivery end, LNG tankers, port facilities, a revaporisation plant and connections to the distribution grid, the complete system capable of supplying $10^9$ scfd of natural gas, the LNG being processed in four streams and shipped over a distance of 3000 miles.

Further assumptions include a price of raw gas into the liquefaction plant of 15 ¢/MM Btu, an overall processing loss of gas of 10%, a storage capacity at the supply end of 100 000 and at the delivery end of 300 000 tons respectively, ships of 125 000 t dwt capable of carrying 75 000 tons of cargo at 17 knots average speed.

This results in the following approximate investments at mid-1972 prices:

|  | MM$ |
|---|---|
| Liquefaction plant and storage | 500 |
| 6 Ships at $75 MM each | 450 |
| Port facilities, jetty, etc. | 20 |
| LNG storage, revaporisation, connections | 20 |
| Contingency, miscellaneous, escalations | 100 |
|  |  |
| Total | 1090 |

150

The effect on delivered gas cost of this investment depends largely on prevailing interest rates and required return on investment. For a project financed one half by fixed interest loans, one half by equity, one can assume an average return of 15%. This results in a delivered gas price of:

|                                                                        | ¢/MM Btu |
| ---------------------------------------------------------------------- | -------- |
| Gas into liquefaction plant                                            | 15       |
| Capital cost (15% p.a. of total investment)                            | 50       |
| Operating cost (manning, fuel, materials, insurance, taxation, utilities, port and other charges, etc.) | 25 |
| Total                                                                  | 90       |

LNG on a year round basis would therefore raise the cost of delivered gas by 90–50 = 40 ¢/MM Btu, an average increase of 10 ¢ for 25% replacement and 20 ¢ for 50% replacement.

If imported LNG is used only during the winter months to meet seasonal gas shortages and shipments are assumed at the same rate but restricted to six months per year, investment remains the same. However, capital costs would become double the original figure and operating cost would also be somewhat higher. The delivered cost of winter only LNG will thus be approximately $1.50/MM Btu, or about $1.00 above the average price of pipeline gas. An average price increase of 25 ¢/MM Btu for 25% penetration and 50 ¢/MM Btu for 50% penetration will be the outcome.

However, since peak load supplies and seasonal gas sales are generally priced much higher than the average pipeline gas, as discussed in Appendix C in greater detail, there remains a considerable incentive to import LNG during the winter, provided supply contracts with LNG producers who have spare capacity can be negotiated.

## FURTHER READING

Bourguet, J. M. (1971). *Oil Gas J.*, **69**(35), 71–75.
Bourget, J. M. (1972). *Oil Gas J.*, **70**(36), 74.

## *Appendix C*

# Optimisation of Peak Shaving Methods

A gas distributor normally purchases his supply from a pipeline under a tariff designed to improve his load factor. This means that the gas price will be made up of a demand charge $d$, expressed in terms of maximum demand per day, and a commodity charge $C$, payable per unit of gas purchased. In addition he will be charged at a much higher rate for excess gas, i.e. for gas supplies over and above $d$ units per day. This excess charge $E$, again per unit of gas purchased, provides the main incentive to look for other sources of supply.

In order to compare the cost of gas supplied by pipeline with that of gas from other sources it is desirable to express the cost of the latter in similar terms, i.e. as a demand charge $d_1$ and a commodity charge $C_1$ although this may not be strictly true where, for instance, an investment in LNG storage enables one to acquire LNG almost at the going gas rate, i.e. without a demand charge. In such a case it is desirable to express the investment $I$ as a capital cost, e.g. in terms of interest and retirement chargeable for a loan, or as a return on the investment plus depreciation. Thus

$$d_1 = \frac{I_1 \text{ (interest rate} + 10\%) + \text{Operating cost}}{\text{Send-out capacity}}$$

assuming that the loan is repayable over ten years. The commodity charge for LNG, of course, remains $C$ but includes an element of conversion efficiency since in order to send out one unit of gas in the winter one may have to purchase somewhat more summer gas to allow for losses, recompression and vaporisation. In other words, all variable costs are grouped together under $C_1$, all fixed charges are included in $d_1$.

The cost of each unit of gas from any one source therefore amounts to

$$T_1 = C_1 + \frac{d_1}{\text{number of days operation } (n_1)}$$

and if gas from more than one source is distributed the total cost of the gas

152

*Sources* [handwritten annotation]

*cost commodity* [handwritten annotation]

$$T = \sum_1^n \left(C_1 + \frac{d_1}{n_1}\right) V_1 \qquad (1)$$

*cost using stored gas* [handwritten annotation]

*include curve* [handwritten annotation]

*fixed → prorate costs to customer on days used* [handwritten annotation]

In order to match a demand/duration curve of the type shown in Chapter 3, i.e. a plot of gas demand vs. the number of days on which it occurs at minimum total cost $T_{min}$ one also has to meet the condition that gas volume for any day of the year, $V$, must equal send out from all available gas sources, i.e.

$$V_{peak} = \Sigma \, V_1 \qquad (2)$$

The cost of gas if a particular form of peak load shaving is in operation for one day is

$$T_1 = (d_1 + C_1) V_1$$

and if the plant is running for $n$ days in a year it becomes

$$T_1 = \left(\frac{d_1}{n} + C_1\right) V_1 \qquad (3)$$

and the number of days in a year when peak shaving system 1 can be used advantageously is defined as

$$T_1 \leqq T \text{ pipeline} \qquad (4)$$

or

$$\frac{D}{n} + C \geqq \frac{d_1}{n_1} + C_1$$

Complications arise from the fact that both $d_1$ and $C_1$ may be functions of the size of the equipment and its capacity. This can be allowed for by considering each size of peak shaving plant separately, always making sure that not only daily capacity but also the total volume of gas stored for peak shaving is sufficient to meet the demand during $n$ days operation.

A simplified table with purely notional gas prices in terms of assumed demand charges and commodity costs is shown overleaf.

From the table it will be gathered that under the circumstances assumed here it will be cheapest to use standard piped gas for all gas volumes required for 100 days or more. To meet additional demand lasting between about 20 and 100 days a year it will pay the utility to liquefy gas in the summer in order to store it and boost winter availability. For peaks between 3 and 15 days duration an LPG supply will be the most economic answer; and the excess demand facilities made available by the supplier should not be used for more than three days a year.

Returning to the total cost of the send-out gas (eqn. 1), this can now be calculated with the aid of the appropriate load/duration

### Average Cost of Send-out Gas
### ($/MM Btu/day)

| | Pipeline gas | | Peak shaving gas | | |
| | Standard | Excess | ex-LNG | LPG | Manu-factured |
|---|---|---|---|---|---|
| Assumed demand charge | 25·00 | nil | 20·00 | 12·00 | 30·00 |
| Commodity cost | 0·50 | 6·00 | 0·55 | 1·00 | 1·00 |

| No. of days | Calculated gas cost vs. days operation | | | | |
|---|---|---|---|---|---|
| 1 | 25·50 | 6·00[a] | 20·55 | 13·00 | 31·00 |
| 2 | 13·00 | 6·00 | 10·55 | 7·00 | 16·00 |
| 4 | 6·75 | 4·50 | 5·55 | 4·00 | 8·50 |
| 8 | 3·62 | 4·25 | 3·05 | 2·50 | 4·75 |
| 16 | 2·06 | 4·13 | 1·80 | 1·75 | 2·87 |
| 32 | 1·28 | 4·06 | 1·18 | 1·38 | 1·94 |
| 64 | 0·89 | 4·03 | 0·86 | 1·19 | 1·47 |
| 100+ | 0·75 | | | | |

[a] Lowest cost supply in heavy type.

curve. First the optimum number of peak shaving days for each system is read from the table. Next the capacity of each system is checked to ensure that it meets maximum demand when in operation and also that it is not unnecessarily large. If a discrepancy is found investment and demand charge values and gas cost are corrected and the optimum number of peak shaving days is checked.

This now enables us to calculate the savings obtainable from the introduction of a specific peak shaving system by comparison with any other peak shaving method or with the piped gas supply. Savings will equal the difference in demand charge multiplied by peak day demand reduced by the volume of peak gas for the entire period of operation, multiplied by the difference in commodity cost, i.e.

$$\text{Savings} = V_{peak}(d_2 - d_1) - \sum_1^n V_1 (C_2 - C_1) \qquad (5)$$

and, clearly, additional peak shaving methods will be introduced only if the value of eqn. (5) is significant.

However, having decided on the methods, estimated the number of days during which each should be operated, obtained the gas volumes for each peak shaving day, and finally introduced the corresponding demand and commodity charges, total gas cost can be calculated in accordance with eqn. (1).

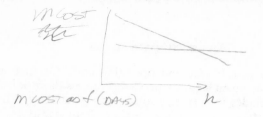

$m \, COST$

$m \, COST \, \propto f \, (DAYS)$   $n$

## Appendix D

# Off-Shore Gas Liquefaction — Economics

The liquefaction of natural gas produced off-shore normally necessitates a system of collecting lines and a platform or well-head to shore pipeline. The gas is piped to a shore-based purification and liquefaction plant, converted into LNG, stored and re-exported by ship from suitable docks or off-shore loading facilities.

As an alternative to this scheme at least one engineering contractor has come up with a proposal for an off-shore liquefaction plant. Rationale for and the general characteristics of such a plant have been discussed in Chapter 3. The present Appendix and particularly Table D.1 are concerned with the economics of such a scheme compared with the economics of a shore-based plant.

### TABLE D.1
#### Economics of Off-shore Gas Liquefaction[a,d]

| | | | |
|---|---|---|---|
| Plant capacity, $10^9$ m³/year | 1·6 | 2·4 | 3·2 |
| Investment estimate, $10^6$ $(1971) | 131 | 152 | 173 |
| Oper. cost, $10^6$ $/month, | 1·9–2·2 | 2·3–2·6 | 2·7–3·0 |
| of which—net oper. cost | 0·26–0·29 | 0·36–0·39 | 0·45–0·48 |
| —capital charges | 1·64–1·91 | 1·94–2·21 | 2·25–2·52 |
| Total liquefaction cost, $/ton | 18–21 | 15–17 | 13–15 |
| Assumed well-head price, $/ton | 5·0 (= about 1·0¢/therm) | | |
| FOB price for LNG, $/ton | 23–26 | 20–22 | 18–20 |
| ¢/therm | 4·5–5·0 | 3·9–4·3 | 3·5–3·9 |
| For comparison: | | | |
| Brunei, first stage, price, ¢/therm | about 3·3[b] | | |
| (Capacity $6 \times 10^9$ m³/year, construction 1970–72) | | | |
| Bechtel Inc. estimate, ¢/therm | 3·5–4·0[c] | | |
| (Capacity $10 \times 10^9$ m³/year, construction 1973) | | | |

[a] Backhaus, H. W. and Zuber, H. (1971). FLOFF, floating gas liquefaction and storage unit for processing off-shore natural gas, *Gas J.*, **348**, 137–153.
[b] Faridany, E. (1972). *LNG Marine Operations*, QER Special No. 12, The Economist Intelligence Unit, London.
[c] Uhl, A. E. and Giese, J. R. (1972). *Economics of LNG Export–Import Systems*, ASME Paper 72-Pet-43, New Orleans.
[d] Backhaus, H. W. (1972). *Model of an LNG Distribution System for North Europe*, Session VII, Paper 6, LNG-3, Washington.

Approximate investments and operating costs for such a plant indicate that investment is somewhat higher, unless shore liquefaction expenditure includes very large sums for gas pipelines and harbour facilities. Using typical figures, however, on-shore investment should be less, particularly for the rather larger plants which are now under construction.

However, operating costs of off-shore plants could be lower owing to:

— the possibility of debt financing;
— the possibility of avoiding local taxation.

If one were to register and establish domicile of a ship carrying a liquefaction plant in a country where taxation was low and if the operator of the off-shore gas concession merely chartered the plant and ship this could result in substantial savings. Credit facilities for ship construction are well established in many countries and maritime tax havens exist in many parts of the world.

Assuming for the moment that such an operation was feasible and that the host government did not raise any objections, then the economics shown in the table might well apply. FOB prices of LNG produced in a floating off-shore liquefaction plant could, under these circumstances, be competitive and plans to build such a plant might go ahead. On the other hand, if plant operations were to be fully taxed and capital returns commensurate with the risk of possible expropriation had to be allowed for, chances of profitable operation of such a plant would definitely recede.

## Appendix E

# Conversion Factors for LNG

The multiplicity of units used to express volumes, weights and heat equivalents of LNG are shown in the following table, which allows one to convert quantities, referring to pure methane, from one set of units into another. Minor corrections should be applied where the composition of the gas is very different, but the factors are of sufficient accuracy down to about 90% $CH_4$.

Conversion Factors for LNG

| | Metric ton liquid | ft³ liquid | m³ liquid | Bbl liquid | Gal liquid | ft³ gas | m³ gas | Million Btu | Million kcal |
|---|---|---|---|---|---|---|---|---|---|
| 1 metric ton liquid | 1 | 84·56 | 2·394 | 15·06 | 632·5 | 52,890 | 1,420 | 52·89 | 13·33 |
| 1 ft³ liquid | 0·01183 | 1 | 0·02831 | 0·1781 | 7·479 | 625·4 | 16·79 | 0·6254 | 0·1576 |
| 1 m³ liquid | 0·4177 | 35·32 | 1 | 6·29 | 265·4 | 22,090 | 593·1 | 22·09 | 5·567 |
| 1 bbl liquid | 0·0664 | 5·615 | 0·1590 | 1 | 42 | 3,512 | 94·27 | 3·512 | 0·8850 |
| 1 gal liquid | 0·001581 | 0·1337 | 0·003786 | 0·02381 | 1 | 83·62 | 2·245 | 0·08362 | 0·02107 |
| 1 ft³ gas $\times$ 10⁶ | 18·91 | 1,599 | 45·27 | 284·8 | 11,960 | 10⁶ | 26,850 | 1,000 | 252 |
| 1 m³ gas $\times$ 10⁶ | 704·4 | 59,560 | 1,686 | 10,610 | 445,400 | 35·32$\times$10⁶ | 10⁶ | 35,320 | 8,900 |
| 1 million Btu | 0·01891 | 1·599 | 0·04527 | 0·2848 | 11·96 | 1,000 | 26·85 | 1 | 0·252 |
| 1 million kcal | 0·07502 | 6·345 | 0·1796 | 1·130 | 47·46 | 3,968 | 112·4 | 3·968 | 1 |

# Natural Gas Liquefaction Plants Around the World

The following tables list available information on the world's natural gas liquefaction plants which are either operating or under construction. We have not included so-called satellite plants, of which there were another 44 in North America alone.

The tables list names of operator and location of the plant, liquefaction, storage and regasification capacity, type of regasification, storage and liquefaction, contractors names for liquefaction and storage, and date of completion.

Peak shaving plants in the United States, peak shaving plants elsewhere and liquefaction plants for the international LNG trade are each listed separately. The most recent available information has been used, but frequent announcements of changes in design or capacity of plants under construction and release of information since going to press must of necessity affect the accuracy of some of the tables.

## NOTES FOR TABLES F.1 to F.4

| | |
|---|---|
| C & BI | Chicago Bridge and Iron |
| PDM | Pittsburgh des Moines Steel |
| APCI | Air Products and Chemicals Inc. |
| Cascade | Conventional multi-refrigerant cascade (Phillips, Pritchard, Air Liquide, Technip) |
| Modified Cascade | Mixed refrigerant (Technip, Pritchard, Linde, Air Liquide) |
| MCR | Multi-component refrigerant (Air Products, Esso R & E)—Propane precooled modified cascade |
| MRL | Mixed refrigerant liquefaction (Chicago Bridge & Iron) |
| PRICO | Poly-refrigerant integral cycle (Pritchard) |
| MRC | Mixed-refrigerant cascade (Cryoplants) |
| Expander | Expander refrigeration cycle (Messer, Chicago Bridge, Cryoplants) |

**TABLE F.1**

**US LNG Peak Shaving Facilities**

| Company and plant site | Liquefaction capacity (MMscfd) | Storage capacity (MMscf) | Regasification | | | Storage container | Contractor | | Type of cycle | Year of operation |
|---|---|---|---|---|---|---|---|---|---|---|
| | | | Design capacity (MMscfd) | Runs | Type | | Storage | Liquefaction plant | | |
| **SAN DIEGO GAS & EL. CO.** Chula Vista, Calif. | | | | | | | | | | |
| 1st plant | 2·0 | 625 | 120 | 2 | Submerged | Above ground | Chicago Bridge | Am. Messer | Expander | 1965 |
| 2nd plant | 7·0 | 1 200 | 180 | 3 | Submerged | 9% Ni | Chicago Bridge | Airco Cryoplants | Expander–nitrogen | 1970 |
| **ALABAMA GAS CO.** Birmingham, Ala. | | | | | | | | | | |
| 1st tank | 4·65 | 625 | 42·5 | 1 | Submerged | Above ground 9% Ni | Chicago Bridge | Air Products | Cascade | 1965 |
| 2nd tank | — | — | — | — | — | Above ground, Al | Chicago Bridge | — | — | 1970 |
| Montgomery, Ala. | 2·7 | 600 | 80 | 2 | Submerged | Above ground, Al | Chicago Bridge | Chicago Bridge | Mod. cascade | 1972 |
| WISCONSIN NAT. GAS CO. Oak Creek, Wis. | 0·75 | 256 | 50 | 2 | Intermed. fluid | Above ground, Al | Chicago Bridge | Chicago Bridge | Cascade | 1965 |
| TRANSCONTINENTAL GAS Hackensack, N.Y. | 7·0 | 1 000 | 200 | 2 | Intermed. fluid | Above ground, Al | Chicago Bridge | J. F. Pritchard | Cascade | 1965 new tank |
| MEMPHIS LIGHT GAS & WATER Memphis, Tenn. | 5·0 | 1 000 | 201 | 3 | Submerged | Above ground, 9% Ni | Chicago Bridge | Air Liquide | Cascade | 1967 |
| APCI–NEGEA Hoplinton, Mass. | 18·5 | 2 000 | 248 | 4 | Submerged | Above ground, 9% Ni | Chicago Bridge | Cyro Methane | Cascade | 1967 new tank |
| **BOSTON GAS CO.** Boston, Mass. | | | | | | | | | | |
| 1st tank | 6·0 | 1 000 | 187·5 | 3 | Submerged | Above ground, 9% Ni | Chicago Bridge | Chicago Bridge | Expander | 1968 |
| 2nd tank | — | 1 120 | — | — | — | Above ground, Al | Chicago Bridge | — | — | 1971 |
| **BROOKLYN UNION GAS CO.** Brooklyn, N.Y. | | | | | | | | | | |
| 1st tank | 5·68 | 625 | 100 | 2 | Direct fired | Above ground, 9% Ni | Chicago Bridge | Am. Messa | Expander | 1968 |
| 2nd tank | — | 1 000 | — | — | — | Above ground, 9% Ni | Chicago Bridge | — | — | 1971 |

| Company / Location | | | | No. | Vaporizer | Storage | | | Process | Year |
| --- | --- | --- | --- | --- | --- | --- | --- | --- | --- | --- |
| NORTHWEST NAT. GAS CO. Portland, Oreg. | 2·5 | 625 | 120 | 2 | Direct fired | Above ground, Al | Chicago Bridge | Chicago Bridge | Expander | 1968 |
| LOWELL GAS CO. Lowell, Mass. | 4·5 | 1 000 | 60 | 3 | Direct fired | Above ground, 9% Ni | Graven/Envirogenics | Envirogenics | Expander | 1969 |
| NORTHERN STATES POWER Eau Claire, Wis. | 2·0 | 270 | 24 | 2 | Direct fired | Above ground, Al | Chicago Bridge | Chicago Bridge | Expander | 1969 |
| FALL RIVER GAS CO. Fall River, Mass. | 0·5 | 150 | 10 | 1 | Direct fired | Above ground, 9% Ni | PDM | CVI Corp. | Direct | 1970 |
| UNION LIGHT, HEAT & POWER CO. Erlanger, Ky. | 1·0 | 16 | Production plant only | | | Above ground, Al | Chicago Bridge | Chicago Bridge | MRL | 1970 |
| PHILADELPHIA GAS WORKS Philadelphia, Pa. | 16·0 | 4 000 | 500 | 5 | Interm. fluid | Above ground, Prestressed Concrete | J. F. Pritchard | J. F. Pritchard | Cascade | 1969 plant / 1972 tanks |
| TEXAS EASTERN TRANS. CO. Staten Island, NY. | 10·3 | 2 040 | 292 | 3 | Direct fired | Above ground, Prestressed Concrete | Texas Eastern | Cryogenics | Mod. cascade | 1970 |
| PHILADELPHIA ELECTRIC CO. Philadelphia, Pa. | 6·6 | 1 200 | 310 | 5 | Submerged | Above ground, 9% Ni | Chicago Bridge | Airco/Cryoplants | Expander/$N_2$ | 1971 |
| LONG ISLAND LIGHTING CO. Holbrook, NY. | 3·0 | 600 | 200 | 4 | ? | Above ground, 9% Ni | PDM | Lotepro Corp. | Cascade/Mixed | 1971 |
| BALTIMORE GAS & EL. CO. Baltimore, Md. | 2·75 | 500 | 250 | 4 | ? | Above ground, Al | Chicago Bridge | Chicago Bridge | Expander | 1971 |
| ATLANTA GAS & LIGHT CO. Riverdale, Ga. — 1st tank | 10·0 | 1 000 | 200 | 3 | Direct fired | Above ground, Al | Chicago Bridge | Air Liquide | Mod. cascade | 1972 |
| | — | 1 500 | — | — | — | Above ground, 9% Ni | Chicago Bridge | — | — | 1972 |
| 2nd tank | 5·0 | 1 000 | 180 | 3 | Submerged | Above ground, Al | PDM | Air Liquide | Mod. cascade | 1972 |
| CITIZENS GAS & COKE Beech Grove, Ind. | 6·6 | 1 200 | 50 | 2 | Vapour transfer | Above ground, Al | PDM | Airco/Cryoplants | Expander/$N_2$ | 1972 |
| CONNECTICUT NAT. GAS CORP. Rocky Hill, Conn. | 5·0 | 1 200 | 105 | 3 | Direct fired | Above ground, 9% Ni | PDM | Envirogenics | Cascade | 1972 |
| COMMONWEALTH NAT. GAS CO. Tidewater, Va. | 1·5 | 250 | 50 | 2 | Direct fired | Above ground, Al | Chicago Bridge | Lotepro | ? | 1972 |
| DELMARVA POWER & LIGHT CO. Wilmington, Del. | 2·0 | 500 | 50 | 2 | Interm. fluid | Above ground, Al | Chicago Bridge | Chicago Bridge | MRL | 1972 |
| IOWA-ILLINOIS GAS & EL. CO. Bottendorf, Ia. | 10·0 | 2 000 | 300 | 3 | Direct fired | Above ground, Al | Chicago Bridge | Chicago Bridge | Expander | 1972 |
| PEOPLES GAS, LIGHT & COKE Mahomet, Ill. | | | | | | | | | | 1972 |
| ROANOKE GAS CO. Roanoke, Va. | 1·0 | 200 | 20 | 2 | Interm. fluid | Above ground, Al | Chicago Bridge | Chicago Bridge | MRL | 1972 |

**TABLE F.1**—*(continued)*

| Company and plant site | Liquefaction capacity (MMscf/d) | Storage capacity (MMscf) | Regasification Design capacity (MMscf/d) | Runs | Type | Storage container | Contractor Storage | Liquefaction plant | Type of cycle | Year of operation |
|---|---|---|---|---|---|---|---|---|---|---|
| ARKANSAS–MISSOURI POWER CO. Blytherville, Ark. | 0·75 | 190 | ? | ? | ? | Above ground, 9% Ni | Brown Minneapolis | CTI | Stirling | 1973 |
| CHATTANOOGA GAS CO. Chattanooga, Tenn. | 10·00 | 1 200 | 60 | 3 | Direct fired | Above ground, 9% Ni | PDM | J. F. Pritchard | ? | 1973 |
| SOUTHERN CONNECTICUT GAS CO. Bridgeport, Conn. | 6·6 | 1 200 | 60 | 2 | Vapour transfer | Above ground, 9% Ni | PDM | Airco/Cryoplants | Expander/$N_2$ | 1973 |
| SPRINGFIELD GAS LIGHT CO. Ludlow, Mass. | 5·0 | 1 000 | 55 | 3 | ? | Above ground 9% Ni | Chicago Bridge | J. F. Pritchard | Mixed refr. | 1973 |
| U.G.I. CORP. Reading, Pa. | ? | 250 | 25 | ? | ? | Above ground, 9% Ni | PDM | Airco/Cryoplants | Expander/$N_2$ | 1973 |
| CONSOLIDATED EDISON CO. Astoria, N.Y. | 6·0 | 1 000 | 300 | 5 | ? | Above ground, 9% Ni | PDM | Airco/Cryoplants | Expander/$N_2$ | 1973 |
| NASHVILLE GAS CO. Nashville, Tenn. | 5·0 | 1 000 | 150 | 3 | Direct fired | Above ground, Al | Chicago Bridge | Chicago Bridge | MRL | 1973 |
| PIEDMONT NAT. GAS CO. Charlotte, N. Carolina | 5·0 | 1 000 | 150 | 3 | Direct fired | Above ground, 9% Ni | Chicago Bridge | Chicago Bridge | MRL | 1973 |
| MASSACHUSETTS LNG INC. (NEES) | 7·85 | 1 000 | 86·4 | 3 | Submerged | Above ground, 9% Ni | PDM | Air Products & Chem. | MCR | 1973 |
| KOKOMO GAS & FUEL CO. Kokomo, Ind. | 1·3 | 400 | 30 | 2 | Direct fired | Above ground, Al | Chicago Bridge | Chicago Bridge | MRL | 1973 |
| INDIANA PUBLIC SERVICE CO. La Porte, Ind. | 10·0 | 2 000 | 300 | 3 | Submerged | Above ground Al | Chicago Bridge | J. F. Pritchard | PRICO | 1974 |
| ALABAMA–TENNESSEE NAT. GAS CO. Green Brier, Ala. | 2·0 | 400 | 30 | ? | ? | ? | ? | J.F. Pritchard | ? | 1974 |
| PACIFIC LIGHTING CO. Cook Inlet, Alaska | 250 | 400 | Production plant only | | | | ? | ? | ? | 1975 |

## TABLE F.2
## LNG Peak Shaving Facilities Outside the US

| Company and plant site | Liquefaction cap. (MMscfd) | Storage cap. (MMscf) | Regasification | | | | Contractor | | Type of cycle | Year of operation |
|---|---|---|---|---|---|---|---|---|---|---|
| | | | Design cap. (MMscfd) | Runs | Type | Storage container | Storage | Liquefaction plant | | |
| NORTHERN & CENTRAL GAS Hagar, Ont. | 2·5 | 625 | 85 | 2 | Submerged | Above ground 9% Ni | Horton (Chic. Bridge) | Air Liquide | Modif. cascade | 1968 |
| BRIT. GAS Ambergate, Derby | 0·5 | 265 | 3·2 | 1 | | Above ground Al | Chicago Bridge Inc. | Whessoe | Helium expander | 1969 |
| GAS METROPOLITAN Montreal, Queb. | 10·0 | 2 000 | 120 | 2 | Submerged | Above ground 9% Ni | Horton (Chic. Bridge) | Air Liquide | Modif. cascade | 1969/72 |
| B.C. HYDRO & POWER Vancouver, BC | 2·5 | 625 | 150 | 3 | Submerged | Above ground 9% Ni | Pittsburgh Des Moines | Air Liquide | Modif. cascade | 1971 |
| TECHN. WERKE Stuttgart, W. Germany | 2·4 | 770 | 63·0 | 3 | Submerged | Above ground 9% Ni | Linde A.G. | Heinrich | Modif. cascade –open cycle | 1971 |
| BRIT. GAS Glenmavis, Scotland I | 5·3 | 1 000 | 260 | 4 | Submerged | Above ground 9% Ni | Motherwell Bridge | Koppers Cryoplants | MRC | 1972 |
| BRIT. GAS Partington, Cheshire I | 10·6 | 2 000 (2 tanks) | 450 | 5 | Submerged | Above ground 9% Ni | Wm. Neill | Cryoplants | MRC | 1973 |
| BRIT. GAS Glenmavis, Scotland II | 5·3 | 1 000 | — | — | — | Above ground 9% Ni | Motherwell Bridge | Cryoplants | MRC | 1974 |
| BRIT. GAS Canvey Island, Essex | 10·6 | — | — | — | — | — | — | Cryoplants | MRC | 1974 |
| NEDERL. GASUNIE Rotterdam, Holland | 12·0 | 2 700a | 430 | ? | ? | Above ground 9% Ni | ? | ? | Ethylene precooled MRC | 1974/5 |
| BRIT. GAS Partington, Cheshire II | | 2 000 (2 tanks) | — | — | — | Above ground Al | Whessoe | — | — | 1974/5 |
| BRIT. GAS Hirwaun, S. Wales | 10·6 | 2 000 | 260 | 4 | Submerged | Above ground 9% Ni | ? | Cryoplants | MRC | 1975 |
| BRIT. GAS Plymouth, Devon | 10·6 | 2 000 | 260 | 4 | Submerged | Above ground 9% Ni | ? | Cryoplants | MRC | 1975 |

a Two 57 000 m³ methane, one 16 000 m³ liquid nitrogen tank.

TABLE F.3

International Natural Gas Liquefaction Facilities

| Company and plant site | Liquefaction capacity (Scfd) | Storage capacity (MM scf) | Type of storage | Storage contractor | Liquefaction plant contractor | Type of cycle | Year of operation |
|---|---|---|---|---|---|---|---|
| CAMEL Arzew, Algeria | 150 | 1 840 | Above ground | Chicago Bridge Licensee | Technip | Cascade | 1963–5 |
| PHILLIPS–MARATHON Kenai, Alaska | 90 | 2 722 | Above ground Al | Chicago Bridge | Phillips | Cascade | 1969 |
| ESSO LIBYA Marsa el Brega, Libya | 345 | 2 100 | Above ground 9% Ni | Chicago Bridge | Air Products–Esso Research | MCR | 1970 |
| SOMALGAS Skikda, Algeria | 360 360 | 2 500 1 250 | Above ground 9% Ni | Chicago Bridge Licensee | Technip–Pritchard-Mandes | Modif. cascade Modif. cascade | 1972/3 1973/4 |
| BRUNEI LNG Lumut, Brunei, Borneo | 500–750 | 2 700 | Above ground 9% Ni | Togo Kanetsu | Japan Gasoline–UOP | MCR | 1972/3 |
| SONATRACH Arzew, Algeria | 1 000 | ? | ? | ? | Procon-Chem. Constr. Co. | ? | 1975/6 |
| SARAWAK LNG Borneo | 870–1 015 | ? | ? | ? | ? | ? | 1975/6 |
| ABU DHABI LIQUEFACTION Abu Dhabi | 435 | ? | ? | ? | Bechtel-Chioda | ? | 1976 |

**TABLE F.4**

**LNG Receiving Terminals**

| Company and plant site | Storage capacity (MMscf) | Regasification Capacity (MMscf/d) | Runs | Type | Type of storage | Storage contractor | Storage design | Year of operation |
|---|---|---|---|---|---|---|---|---|
| BRITISH GAS COUNCIL Canvey Island, Essex | 4 000 (4 tanks) 1 400 (8) | 100 | 4 | Direct fluid | Frozen hole Above ground, Al | Chicago Bridge Licensee | Gas Council | 1958-67 |
| GAZ DE FRANCE Le Havre, France | 792(3) | 70 | 2 | Direct fluid | Above ground 9% Ni | Chicago Bridge Licensee | Gaz de France | 1963 |
| GAS NATURAL Barcelona, Spain | 1 800(2) | 110 | 2 | Intermed. fluid | Prestressed concrete | Preload Co. | Esso R. & E. | 1969 |
| | 1 800 | — | — | — | Above ground 9% Ni | Chicago Bridge Licensee | C. B.& I. | 1973 |
| SNAM La Spezia, Italy | 2 100(2) | 235 | — | Intermed. fluid | Above ground 9% Ni | SAIPEM (PDM Licensee) | Esso R. & E. | 1970 |
| TOKYO GAS/ELECTRIC Nagishi, Japan | 3 300(5) | 220 | — | Direct | Above ground 9% Ni | I.H.I. (Chicago Bridge Lic.) | I.H.I. | 1969 |
| DISTRIGAS CORP. Everett, Mass. | 3 250(2) 6 000 | 135 — | 3 | Submerged ? | Above ground 9% Ni Prestressed concrete | Chicago Bridge Preload/Walsh | C.B. & I. — | 1971 1973 |
| COLUMBIA LNG Cove Point, Maryland | 5 000 | 1 000 | ? | ? | Above ground | ? | ? | 1975 |
| GAZ DE FRANCE Fos-sur-Mer, France | 1 500(2) | ? | ? | ? | Above ground, Al | Chicago Bridge Lic. | Gaz de France | 1971 |
| OSAKA GAS CO. Osaka, Japan | 3 000(3) | — | — | — | Above ground, Al | I.H.I. (Chicago Bridge Lic.) | I.H.I. | 1973 |
| PACIFIC LIGHTING Los Angeles, Cal. | 4 000(3) | ? | ? | ? | ? | ? | ? | 1975 |
| SOUTHERN ENERGY CO. Savanah, Georgia | 4 000(2) | ? | ? | ? | Above ground | ? | ? | 1975 |
| MONFALCONE Trieste, Italy | ? | ? | ? | ? | Above ground | ? | ? | 1975 |

In addition there are some 50 small satellite LNG plants in the US and elsewhere which receive liquid LNG, generally by road, and have regasification facilities.

**TABLE F.5**

**International Gas Liquefaction Projects of the Future**

| Exporting area | Plant location | Proposed by | LNG destination | Volume (MMscfd) | Year | |
|---|---|---|---|---|---|---|
| Algeria | Skikda | El Paso | US | 2 000 | 1976 | |
| | Delliah | Eascogas | US | 920 | 1975/6 | some delay expected |
| | Delliah | Sonatrach | France, Belgium Germany | 1 000 | 1977/8 | |
| Nigeria | Skikda | Gas Natural | Spain | 30–150 | 1976 | |
| | Port Harcourt | Shell–B.P. | US | 650 | ? | |
| | ? | Guadalupe Gas/Nig. Gov. | US | ? | ? | |
| | ? | Gulf Oil | US | 500 | 1980 | |
| Middle East | Qatar | Qatar Petroleum | Japan | 120 | 1978 | |
| | Iran (Qeshm) | NIGC/Fuji | Japan | 580–870 | 1977/8 | some doubt |
| | Iran (Kalingas) | NIGC/Int. Syst. etc. | Japan | 800–1 200 | ? | |
| Australasia | Sarawak | Shell/Mitsubishi/Tokyo El | Japan | 1 015 | 1976 | |
| | Palm Valley | Magellan/Flinders | Japan | 750–1 000 | ? | |
| | Palm Valley | Magellan | US | 500 | ? | |
| | Rankin Off-shore | Burmah–Woodside | Japan | 600–1 000 | 1976/7 | mainly depending on Australian Federal Govt. approval |
| | Rankin Off-shore | Burmah–Woodside | US | 500 | ? | |
| | W. Pakistan (Sui) | Burmah | Japan | 500 | ? | |
| | Bangladesh | Government | Japan | ? | ? | |
| | N. Sumatra (Arun) | Pertamina/Mobil | Japan | 1 000 | 1978 | |
| | Indonesia | Pertamina/Pacific Lighting | US | 1 000 | 1978 | some doubt |
| USSR | East Siberia | Governments/El Paso | Japan | 500 | 1980 | |
| | Baltic Port | USSR Govt./Tenneco | US | 2 000 | 1980 | |
| | East Siberia | USSR Govt./El Paso | US | 500 | 1980 | |
| South America | Lake Maracaibo | Cia Venez. de Petroleos | US | 600 | 1976 | doubtful |
| | Ecuador | ADA Oil etc. | US | 500 | ? | doubtful |
| | Trinidad | Amoco/Peoples Gas/T. Govt. | US | 200–400 | 1976 | doubtful |

## Appendix G

# LNG Carriers—Present and Future

The number of ships suitable for the transport of cryogenic cargoes and specifically of LNG has been very small until quite recently. After the construction of two prototypes, the *Methane Pioneer*, subsequently re-named *Aristotle*, and the *Pythagore*, three full scale carriers were built for the Algerian–UK/France trade. These were followed by four more ships to carry gas from Libya to Italy and Spain and two ships for the Alaska–Japan run. The first tanker to be built without a specific operational target was the *Descartes*, mainly intended for freelance operations from North Africa to the US (see Table G.1).

Of the ships now under construction and on order a substantial number may well operate in similar fashion. As shown in Tables G.2, which lists LNG carriers now building, and G.3, which covers those on order but not yet started, only seven ships ordered by the Shell Group to carry LNG from Brunei or Sarawak to Japan and one French tanker have firm engagements; the remainder will have to negotiate charters or will ultimately be sold to the operators of future LNG schemes around the world.

However, the remarkable total of 24 ships under construction and 63 on order at the beginning of 1973 is an indication of the confidence shown by shipowners in the rapid development of the LNG trade, not too unreasonable an assumption if one bears in mind that total exports of LNG from Algeria alone which are now under consideration will require about 30 ships of 125 000 m$^3$ each.[1]

### REFERENCES

1. Ait-Laussine, N. (1972). *Development of LNG in Algeria*, Session III, Paper 7, LNG-3, Washington.
2. Faridany, E. (1972). *LNG Marine Operations*, QER Special No.12, The Economist Intelligence Unit, London.

TABLE G.1

LNG Carriers in Operation (September 1972)[2]

| Name | Owner | Run | Year built | Size ($m^3$) | Tankage |
|---|---|---|---|---|---|
| Aristotle | Antartic Gas | Prototype | 1959 | 5 100 | Prismatic, Free, Aluminium |
| Pythagore | Gaz Ocean | Prototype | 1964 | 630 | SS Membrane (Technigaz), Integral |
| Methane Princess | Conch Methane | Algeria–UK | 1964 | 27 400 | Prismatic, Free, Aluminium |
| Methane Progress | Conch Methane | Algeria–UK | 1964 | 27 400 | Prismatic, Free, Aluminium |
| Jules Verne | Gaz Marine | Algeria–France | 1965 | 25 500 | Cylindrical, Free, 9% Ni steel |
| Laieta | Nav. de Prod. Lic. | Libya–Spain | 1969 | 40 600 | Prismatic, Free, Aluminium |
| Esso Brega | Prora Trasporti | Libya–Italy | 1969 | 40 600 | Prismatic, Free, Aluminium |
| Esso Portovenere | Prora Trasporti | Libya–Italy | 1969 | 40 600 | Prismatic, Free, Aluminium |
| Polar Alaska | Polar LNG Shipping | Alaska–Japan | 1969 | 71 500 | Invar Membrane (Gaz Transport) Integral |
| Arctic Tokyo | Arctic LNG Transp. | Alaska–Japan | 1969 | 71 500 | Invar Membrane (Gaz Transport) Integral |
| Esso Liguria | Prora Trasporti | Libya–Italy | 1970 | 40 600 | Prismatic, Free, Aluminium |
| Euclides | Gaz Ocean | Prototype | 1970 | 4 000 | Spherical (Technigaz), Free, 9% Ni steel |
| Descartes | Gaz Ocean/Sonatrach | Algeria–US | 1971 | 50 000 | SS Membrane (Technigaz) Integral |

## TABLE G.2
## LNG Carriers under Construction (Sept. 1972)[2]

| Name | Owner | Run | Delivery date | Size | Tankage |
|---|---|---|---|---|---|
| Gadinia[a] | Shell Tankers—UK | Brunei-Japan | 1972 | 75 000 | SS Membrane (Technigaz), Integral |
| Gadila[a] | Shell Tankers—UK | Brunei-Japan | 1972 | 75 000 | SS Membrane (Technigaz), Integral |
| Gari[a] | Shell Tankers—UK | Brunei-Japan | 1972 | 75 000 | SS Membrane (Technigaz), Integral |
| Gaztrana[a] | Shell Tankers—UK | Brunei-Japan | 1973 | 75 000 | SS Membrane (Technigaz), Integral |
| M176 | Smedvig—Norway | ? | 1973 | 29 000 | Spherical (Moss), Free, 9% Ni steel |
| Norman Lady | Buries-Markes—UK | Abu Dhabi-Japan | 1973 | 87 600 | Spherical (Moss), Free, 9% Ni steel |
| Charles Tellier | Gazocean/Mess. Marit.–France | Algeria–France (Fos) | 1973 | 40 000 | Invar Membrane (Gaz Transport) Integral |
| R197 | LNG Carriers Ltd—UK | ? | 1974 | 87 600 | Spherical (Moss), Free, 9% Ni steel |
| M177 | Hilmar Reksten—Norway | ? | 1974 | 29 000 | Spherical (Moss), Free, 9% Ni steel |
| 1401 | Lofoten—Norway | ? | 1974 | 35 000 | Spherical (Moss), Free, 9% Ni steel |
| Benjamin Franklin | Gazocean/Armement—France | Algeria–US | 1974 | 120 000 | SS Membrane (Technigaz), Integral |
| ? | Gazocean/Houlder F—UK | ? | 1974 | 88 000 | SS Membrane (Technigaz), Integral |
| DK283 | El Paso Nat. Gas—US | Algeria–US | 1974 | 125 000 | SS Membrane (Technigaz), Integral |
| DK284 | El Paso Nat. Gas—US | Algeria–US | 1975 | 125 000 | SS Membrane (Technigaz), Integral |
| DK287 | El Paso Nat. Gas—US | Algeria–US | 1976 | 125 000 | SS Membrane (Technigaz), Integral |
| Gouldia | Shell Tankers—UK | Brunei-Japan | 1975 | 75 000 | Invar Membrane (Gaz Transport) Integral |
| Gena | Shell Tankers—UK | Brunei-Japan | 1975 | 75 000 | Invar Membrane (Gaz Transport) Integral |
| Genota | Shell Tankers—UK | Brunei-Japan | 1975 | 75 000 | Invar Membrane (Gaz Transport) Integral |
| R198 | Gotaas-Larsen—Norway | Abu Dhabi-Japan | 1975 | 125 000 | Spherical (Moss), Free, 9% Ni steel |
| R200 | Internat. Gas—Norway | ? | 1977 | 87 600 | Spherical (Moss), Free, 9% Ni steel |
| 1402 | Lofoten—Norway | ? | 1975 | 35 000 | Invar Membrane (Gaz Trans.) Integral |
| 302 | Gazocean/Armement—France | ? | 1976 | 120 000 | SS Membrane (Technigaz) Integral |
| A26 | Zodiac Shipping—US | ? | 1976 | 120 000 | SS Membrane (Technigaz) Integral |
| B26 | Ocean Steam Ship—UK | ? | 1976 | 120 000 | SS Membrane (Technigaz) Integral |

## TABLE G.3
### LNG Carriers on Order (Sept. 1972)[2]

| Numbers | Owner | For delivery | Size | Tankage |
|---|---|---|---|---|
| 1 | Esso—US | 1975 | 125 000 | ? |
| 5 | Onassis—Greece | 1975 and later | 165 000 | ? |
| 2 | Buries-Markes/Hoegh—UK/Norway | 1975/76 | 120 000 | ? |
| 6 | El Paso Nat. Gas—US | 1976 | 125 000 | Spherical (Moss), Free, 9% Ni steel |
| 3 | Globtik—UK | 1976 | 125 000 | Spherical (Moss), Free, 9% Ni steel |
| 1 | Gotaas-Larsen—Norway | 1976 | 125 000 | Spherical (Moss), Free, 9% Ni steel |
| 7 | Zapata-Norness—US | 1976 | 125 000 | Spherical (Moss), Free, 9% Ni steel |
| 5 | ERAP/Nat. Iran/Itoh | 1976 | 120 000 | ? |
| 1 | Island Nav. Corp.—US | 1976 | 125 000 | ? |
| 3 | Peoples Gas Co.—US | 1976 | 80 000 | ? |
| 2 | Buries-Markes/Hoegh—UK/Norway | 1976/77 | 120 000 | ? |
| 4 | Globtik—UK | 1977 | 125 000 | ? |
| 5 | Fuji Oil/Marubeni—Japan | 1977 | ? | ? |
| 5 | Intercontinental Gas—Norway | 1977 | 12 000 | ? |
| 2–3 | Shell Tankers—UK | ? | 125 000 | ? |
| 3 | Transport Tech.—US | ? | 125 000 | ? |
| 4 | US Lines—US | ? | 120 000 | ? |
| 3 | Comp. Venez. de Petroleo—Venezuela | ? | 120 000 | ? |

*N.B.*—All ships to operate between Alaska and the other 48 states and also some others for US-destined cargoes will be built in US yards; US government subsidies are available for this purpose.

## TABLE G.4
**Approximate Shipping Distances for Various LNG Projects**
**Nautical Miles**

| Producing country | Distance to LNG project (nautical miles) | | | |
|---|---|---|---|---|
| | US East Coast | US West Coast | English Channel | Japan |
| Venezuela | 1 700 | 3 000ᵃ | 3 500 | — |
| Alaska | — | 17000 | — | 1 800 |
| Trinidad | 1 800 | — | 3 500 | — |
| Ecuador | 3 000ᵃ | 3 200 | — | 6 000 |
| USSR | 4 300 | 4 600 | — | — |
| Libya | 4 100 | — | 2 200 | — |
| Algeria | 3 600 | — | 1 700 | — |
| Nigeria | 5 000 | — | 4 100 | — |
| Brunei | 7 100 | 6 500 | — | 2 400 |
| Australia | 7 000 | 7 400 | — | 3 100 |
| Persian Gulf | 11 700 | 11 000 | — | 6 200 |

ᵃUsing Panama Canal.

# Physical Properties of Principal LNG Components

Natural gases, in addition to methane, contain a number of minor components, the physical properties of which affect the characteristics of the liquefied blend, although concentrations of the individual impurities rarely exceed 10% by volume and frequently amount to less than 1%. The significance of these other components must therefore not be underrated, particularly as far as their total effect is concerned, and it would be totally misleading to consider LNG to all intents and purposes to be liquefied methane.

The following table lists molecular weight, melting point, boiling point, specific gravity, gas density, critical conditions, heats of combustion, inflammability limits, heats of vaporisation, specific heats and vapour pressure of the following gases:

Methane — Ethane — Propane — Isobutane
n-Butane — Isopentane — n-Pentane — Ethylene —
Nitrogen — Carbon Dioxide.

## TABLE H.1
### Selected Physical Constants of LNG Constituents

| Property | Unit | Methane | Ethane | Propane | Butanes (iso) | Butanes (normal) | Pentanes (iso) | Pentanes (normal) | Ethylene | Nitrogen | CO2 |
|---|---|---|---|---|---|---|---|---|---|---|---|
| Molecular weight | | 16.04 | 30.07 | 44.09 | 58.12 | 58.12 | 72.15 | 72.15 | 28.05 | 28.02 | 44.01 |
| Freezing point (1 atm abs) | °C | -182.5 | -183.3 | -187.7 | -159.6 | -138.3 | -159.9 | -129.7 | -169.1 | -209.9 | b |
| | °F | -296.5 | -297.9 | -305.8 | -255.3 | -217.0 | -255.8 | -201.5 | -272.5 | -345.6 | b |
| Boiling point (1 atm abs) | °C | -161.6 | -88.6 | -42.1 | -11.7 | -0.5 | -27.8 | 36.1 | -103.6 | -195.8 | 42.9 |
| | °F | -258.7 | -127.5 | -43.7 | 10.9 | 31.1 | 82.1 | 96.9 | -154.7 | -320.0 | 109.3 |
| Density of liquid (at 1 atm abs) | S.G. 15/15 | 0.3 | 0.3771 | 0.5077 | 0.5631 | 0.5844 | 0.6248 | 0.6312 | 0.35 | — | 0.8159 |
| | °API | 340.0 | 243.7 | 147.2 | 119.8 | 110.6 | 95.0 | 92.7 | — | — | — |
| Density of gas (5°C, 1 atm abs) | Air = 1[a] | 0.554 | 1.038 | 1.522 | 2.066 | 2.066 | 2.491 | 2.491 | 0.968 | 0.967 | 1.5194 |
| | lb/1 000 ft$^3$ [a] | 42.27 | 79.23 | 116.19 | 153.15 | 153.15 | 190.11 | 190.11 | 73.92 | 73.90 | 116.70 |
| | g/m$^3$ | 716.8 | 1 356.0 | 2 019.0 | 2 668.0 | 2 703.0 | 3 195.5 | 3 216.5 | 1 260.5 | 1 250.6 | 1 976.8 |
| Ratio gas vol./liquid vol. (5°C, 1 atm abs) | | 442.4 | 296.8 | 227.5 | 229.3 | 238.0 | 205.0 | 207.0 | — | — | 760.0 |
| Critical temp. | °C | -82.1 | 32.4 | 96.8 | 135.0 | 152.0 | 187.2 | 196.4 | 9.9 | -147 | 31.1 |
| | °F | -115.8 | 90.3 | 206.3 | 275.0 | 305.6 | 369.0 | 385.5 | 49.8 | -232.8 | 88.0 |
| Critical pressure | atm | 45.8 | 48.3 | 42.0 | 36.0 | 37.5 | 32.9 | 33.3 | 50.0 | 33.6 | 73.0 |
| | psia | 673 | 710 | 617 | 529 | 551 | 483 | 490 | 742 | 492 | 1 073 |
| Heat of combustion (at 15°C, gross) | Btu/scf[a] | 1 010 | 1 769 | 2 517 | 3 253 | 3 262 | 4 000 | 4 009 | 1 599 | — | — |
| | kcal/Nm$^3$ [a] | 9 643 | 16 890 | 24 032 | 31 060 | 31 145 | 38 192 | 38 278 | 15 267 | — | — |
| | Btu/lb[a] | 23 885 | 22 323 | 21 664 | 21 238 | 21 299 | 21 041 | 21 088 | 21 834 | — | — |
| | kcal/kg | 13 271 | 12 403 | 12 037 | 11 800 | 11 834 | 11 690 | 11 716 | 12 130 | — | — |
| Inflammability limit (% in air) | lower | 5.0 | 2.9 | 2.1 | 1.8 | 1.8 | 1.4 | 1.4 | 2.7 | — | — |
| | upper | 15.0 | 13.0 | 9.5 | 8.4 | 8.3 | 8.3 | 8.3 | 34.0 | — | — |
| Heat of vapours— (at atm boiling point) | Btu/lb | 219.2 | 210.4 | 183.1 | 157.5 | 165.7 | 147.1 | 153.6 | 207.6 | 86 | 248b |
| | kcal/kg | 121.8 | 116.9 | 101.7 | 87.5 | 92.1 | 81.7 | 85.4 | 115.3 | 47.8 | 137.8 |
| Specific heat (at 15°C, 1 atm) | $C_p$ gas[a] | 0.5271 | 0.4027 | 0.3885 | 0.3872 | 0.3908 | 0.3827 | 0.3883 | 0.3622 | 0.2482 | 0.1991 |
| | $C_v$ gas[a] | 0.403 | 0.344 | 0.343 | 0.353 | 0.357 | 0.355 | 0.361 | 0.291 | 0.177 | 0.153 |
| (at atm, boiling point) | $C_p$ liquid | 0.925 | 0.926 | 0.592 | 0.570 | 0.564 | 0.535 | 0.542 | — | — | — |

[a] Assumes ideal gas behaviour.
[b] Undergoes sublimation.

# Index